Zero

Zero

A Landmark Discovery, the Dreadful Void, and the Ultimate Mind

Syamal K. Sen
GVP - Prof. V. Lakshmikantham Institute for Advanced Studies
GVP College of Engineering Campus
Madhurawada, Visakhapatnam, India

Ravi P. Agarwal
Department of Mathematics
Texas A&M University–Kingsville
Kingsville, TX, USA

ELSEVIER

AMSTERDAM • BOSTON • HEIDELBERG • LONDON
NEW YORK• OXFORD • PARIS • SAN DIEGO
SAN FRANCISCO • SINGAPORE • SYDNEY • TOKYO
Academic Press is an imprint of Elsevier

AP

Academic Press is an imprint of Elsevier
125, London Wall, EC2Y 5AS.
525 B Street, Suite 1800, San Diego, CA 92101-4495, USA
225 Wyman Street, Waltham, MA 02451, USA
The Boulevard, Langford Lane, Kidlington, Oxford OX5 1GB, UK

ISBN: 978-0-08-100774-7

British Library Cataloguing-in-Publication Data
A catalogue record for this book is available from the British Library.

Library of Congress Cataloging-in-Publication Data
A catalog record for this book is available from the Library of Congress.

For Information on all Academic Press publications
visit our website at http://store.elsevier.com/

Contents

Preface

Charles Seife, an American science journalist, writes in 2000 about zero and infinity in his book *Zero*: *The Biography of a Dangerous Idea*: "… They are equally paradoxical and troubling. The biggest questions in science and religion are about nothingness and eternity, the void and the infinite, zero and infinity. The clashes over zero were the battles that shook the foundations of philosophy, of science, of mathematics, and of religion. Underneath every revolution lay a zero—and an infinity." He continues: "… the Greeks banned it, the Hindus worshiped it, and the Church used it to fend off heretics. Now it threatens the foundations of modern physics. For centuries the power of zero savored of the demonic; once harnessed, it became the most important tool in mathematics. For zero, infinity's twin, is not like other numbers. It is both nothing and everything. … Zero has pitted East against West and faith against reason, and its intransigence persists in the dark core of a black hole and the brilliant flash of the Big Bang. Today, zero lies at the heart of one of the biggest scientific controversies of all time: the quest for a theory of everything."

While analogy may be sometimes criticized because two (or more) statements in two different contexts, though may have one-to-one correspondence, could be different significantly. Such an analogy is not only inappropriate but also misleading and one should refrain from using such an analogy. However, in this context we still dare to bring the following analogy.

"There are two parts to a religion—theology and spirituality. There is no difference in spirituality whereas the theology can have many religious dogmas. If we overcome the dogmas and rise to spirituality, there will be no conflict." writes A.P.J. Abdul Kalam (1931–2015), former president of India (2002–2007) in the book *Manifesting Inherent Perfection*: *Education for Complete Self-development* (Sri Ramakrishna Math, Chennai, 2014).

There are analogously two parts of "zero" too—practical usage and spirituality. There is no difference in spirituality whereas practical usage part did have several usages over centuries in different countries finally of course resulting in one globally accepted set of usages. "Zero—a landmark discovery" refers primarily to the practical usage part while "Zero—the dreadful void and the ultimate mind" constitutes the spiritual aspect, rather the highest spiritual point (goal) when one reaches the ultimate mind (state of Samadhi/Silence/No-thought condition implying the complete control over mind—the most difficult task for a common human being)—this "one" is the greatest/wisest living being in the world, nay in the Universe! No living being anywhere in the universe can ever be greater than him!

Seife follows this innocent-looking number (zero) from its birth as an Eastern philosophical concept to its struggle for acceptance in Europe, its rise and transcendence in the West, and its ever-present threat to modern physics. Here are the legendary thinkers—from Pythagoras to Newton to Heisenberg, from the Kabalists to today's astrophysicists—who have tried to understand it and whose clashes shook the foundations of philosophy, science, mathematics, and religion.

We aim at bringing the details of this struggle and the consequent development to light. This monograph records one of the most remarkable discoveries called "zero" both in conventional mathematics as well as in computational mathematics with special reference to natural mathematics (mathematics that nature continuously performs completely error-freely, non-chaotically, and parallely over true/exact real numbers which are, in general, completely out of bound of any man-made digital computer of the past, the present, and also the future). We examine contemporary events occurring side by side in different countries or cultures, reflecting some of the noblest thoughts of generations concerning zero. We document the winding path of its development finally resulting in the Indian zero which is accepted by one and all for all human activities for several centuries. This zero continues to go strong with no further development/improvement and perhaps there will not be one in any foreseeable future. Besides this mundane aspect of zero, there is a much more profound implication of zero in the spiritual plane. We have also explored this aspect with true incidences occurring in recent past.

Certainly a book of this type cannot be written without deriving many valuable ideas from several sources. We express our indebtedness to all authors, too numerous to acknowledge individually, from whose specialized knowledge we have been benefitted. We have also been immensely benefitted from several websites such as en.wikipedia.org as well as from comments specifically by Manas Chanda, former professor of Indian Institute of Science, Bangalore. Special thanks are due to our wives Ella Sen and Sadhana Agarwal whose continued encouragement and sacrifice deserve special mention.

Syamal K. Sen
Visakhapatnam, AP, India

Ravi P. Agarwal
Kingsville, TX, USA

Introduction

1

One of the remarkable things about the behaviour of the world is how it seems to be grounded in mathematics to a quite extraordinary degree of accuracy. The more we understand about the physical world, and the deeper we probe into the laws of nature, the more it seems as though the physical world almost evaporates and we are left only with mathematics.

—Roger Penrose (born 1931).

...With his eyes open, he (Swami Vivekananda, 1863–1902) saw the walls and everything in the room, nay, the whole universe and himself within it, whirling and vanishing into an all-encompassing void. He was frightened as he thought he might be on the verge of death, and cried out: "What are you (Sri Ramakrishna, 1836–1886) doing to me? I have my parents at home."

—Mahendra Nath Dutta (younger brother of Swami Vivekananda). The incidence occurred in November, 1881 in Kolkata.

Zero indicates the absence of a quantity or a magnitude. It is so deeply rooted in our psyche today that nobody will possibly ask "What is zero?" From the beginning of the very creation of life, the feeling of the lack of something or the vision of emptiness/void has been embedded by the creator in all living beings. While recognizing different things as well as the absence of one of these things are easy, it is not so easy to fathom the complete nothingness, viz. the universal void. Although we have a very good understanding of nothingness or, equivalently, a zero today, our forefathers had devoted countless hours and arrived at the representation and integration of zero and its compatibility not only with all nonzero numbers but also with all conceivable environments only after many painstaking centuries. Zero can be viewed/perceived in two distinct forms: (i) as a number in our mundane affairs and (ii) as the horrific void or Absolute Reality in the spiritual plane/the ultimate state of mind. Presented are the reasons why zero is a landmark discovery and why it has the potential to conjure up in an intense thinker the dreadful nothingness unlike those of other numbers such as 1, 2, and 3. Described are the representation of zero and its history including its deeper understanding via calculus, its occurrences and various roles in different countries as well as in sciences/engineering along with a stress on the Indian zero that is accepted as the time-invariant unique absolute zero. This is followed by the significant distinction between mathematics and computational mathematics and the concerned differences between the unique absolute zero and nonunique relative numerical zeros, and their impact and importance in computations on a digital computer.

Zero: A landmark discovery, the dreadful void, and the ultimate mind. DOI: http://dx.doi.org/10.1016/B978-0-08-100774-7.00001-6

1.1 Matter versus nonmatter

While dealing with zero meaning "nothing" or void, its significance in the realm of both matter and nonmatter, its birth and properties, abstract (symbolic) presentation and various names in different contexts, occurrences and uses in science and engineering as well as in different countries are discussed, along with the reasons why it can be portrayed as the most fearful void, the highest state of mind, and also considered as one of the greatest innovations of mankind.

With our current conditioned mind, it appears to us easy to conceive the physical significance of just a zero. It is not difficult to imagine "nothing" in the background of something. It is, on the other hand, very difficult or even dreadful to think "nothing" in the background of "nothing" (achieved by eliminating everything including even the background). Just attempt to think/imagine about something that exists and remove that thing. Continuing the successive removal of one thing after the other and reaching the state in which everything including all relations, the surroundings, and even one's own body from the conscious state of one's mind has vanished, could lead one to a dreadful experience! Was there any universal void—a situation when nothing existed in the Universe? Physics has been sticking until today, and possibly will continue to stick to the point for an indefinite period of time, that something cannot be created from nothing. In other words, nothing can be created out of nothing. This implies there is always something eternally. That is, infinite years ago there was something (matter including energy, assuming that matter is convertible to energy and vice versa), this exists today, and will continue to exist infinite years hence (its form, however, may be changing with time).

In any science including physics, something cannot be created out of nothing. There is no evidence that a thing has been created from complete void. The valiant effort of *Fred Hoyle* (1915–2001 AD) and *Jayant Vishnu Narlikar* (born 1938 AD) during the early 1960s to propound *the Steady State Theory* in Cosmology (an alternative to *the Big Bang Theory* of the universe's origin), which says that new matter is continuously created as the universe expands, thus adhering to the cosmological principle, did not succeed and the theory is now obsolete. This is true for both matter and nonmatter. The mind of any one individual contains all the knowledge (nonmatter). There exists no knowledge outside the mind. It is the specific knowledge-mining that a scientist does in the ocean of knowledge residing in his/her own mind.

According to today's physics, the void, that is, the universal void, was never there, is not there, and will never be there. Two aspects are important to be considered here. One aspect is that of matter while the other aspect is that of nonmatter or, may be termed, Spirit or Nature or God (encompassing all knowledge) or Consciousness that is omnipotent (having unlimited power), omnipresent (present everywhere), and omniscient (knowing everything). One's realization/experience is the proof of the existence of spirit, which is the best proof (better than even a mathematical proof). This Nature (or God if you wish to call it) pervades all matter, all spaces containing matter of varying density including numerically zero (not exactly zero) density. Matter with exactly zero density, that is, completely/absolutely empty space, does not seem to be fathomable by a physicist or possibly by anybody within the realm of science that we are taught conventionally and traditionally. Is there a sharp boundary (maybe static or dynamic) just beyond which matter has absolutely zero density and just within (maybe closest to the boundary) which it has nonzero density? Is there a discontinuity of density (in a mathematical term)? Interestingly, when we attempt to create vacuum in a container, we successively reduce the density of the gas (say, air) but we will never be able to make the density exactly zero by any process that we know of in physics.

1.2 Zero in universal nothingness

Under these circumstances, the zero—the way we understand it today—is distinctly different from other numbers such as 1, 2, 3, and 4 (denoting one, two, three, and four physical objects), which can be very easily comprehended from the physical world which we live in. In this context, we may consider zero, that is, nothingness, in the well-known environment/surrounding of many things which we live with. This zero is well within our understanding but has been playing hide-and-seek over centuries in terms of unambiguous unique representation as well as unambiguous integration with other nonzero numbers (mainly for arithmetic operations) and complete compatibility with everything under all circumstances. *But the zero in the environment of complete vacuum state or absolute nothingness is not well within our grasp.* We therefore stick to the former zero in most of our following discussion.

1.3 Birth and five properties of zero

The exact date of birth of zero is not known although the very feeling of nothingness or of absence (of something) did exist in the minds of living beings since time immemorial. This nothingness is conceived against the visible world around us. The question of uniquely representing this nothingness and its function in relation to other numbers (representing nonnothingness), such as 1, 2, 3, and 4, under all circumstances and in all sciences without any noncompatibility, which has no inner contradiction or clash and which solves all our arithmetic and algebraic problems without any ambiguity, continued to remain elusive to mathematicians for centuries. Today we are so accustomed/conditioned with using zero (0) along with other numbers that we, with our existing mental set-up, will not ask the aforementioned question in the realm of not only arithmetic and algebra but also in the whole of mathematics. For instance, when one subtracts the number 825 from 825, the result is nothing and so an accountant in a business transaction used to keep the result-space blank indicating "nothing." Among a large number of computations, leaving the result-space empty could mean either (i) the accountant has forgotten (a nontrivial possibility) to write the result of the arithmetic expression involving several numbers or (ii) the result of the expression is "nothing" or zero. With our present day conditioned mind it might appear to us that this is not a serious issue as we would readily fill the result-space by one or more zeros. This is a role of zero as a number. Determining (or finding) a symbol for zero different from all other existing symbols was also an issue that might appear trivial to us today, but it was not so during the third or earlier millennium BC. Since zero is the bottom of all positive numbers, it should act as a direction separator to accommodate negative numbers which are unavoidable almost everywhere in science and engineering. In addition, to denote the magnitude of a quantity, a number is used. If the magnitude happens to be nil (that might occur quite often in our physical world, for instance no money or no cow), then the same zero should represent that magnitude. In the Indo-Arabic number system, zero should also act as the place holder. For example, 1 in the unit position and 1 in the tens position are completely different. Adding a zero on the right side of 1 would uniquely decide the value. These five problems did not exist with other nonzero numbers occurring in any arithmetic/mathematical computation that does not encounter zero or "nothing." Thus we should define and represent a zero which have all the foregoing *five* properties. Such a zero has been found to be (would then be) usable everywhere without any context dependence and any ambiguity. There appears to be no other distinct property (besides the foregoing five) that must be satisfied for absolute compatibility with numbers and nonnumbers in any context.

Since the exact date of birth of zero, rather the physical meaning of zero, is unknown and will never be known, one could imagine that zero existed eternally, that is, before the universe (if it is assumed born out of a birthless (visible or non-visible, perceivable or nonperceivable) seed) came into existence and will remain after the universe is gone, like the number Pi (ratio of the circumference and the diameter of any circle or, in other words, the area of the circle with unit radius), but with much more pervasiveness. A primitive/prehistoric man can easily comprehend the absence of something in the background of things around. Thus the concept of zero has been in-built in any primitive man and possibly in any living being from the very beginning of creation of life in the universe. The exact date of birth of the very first primitive man is not known, we can only attempt, based on some controversial logic/reasoning, the approximate large period of time that might contain the exact date of birth of the first primitive man. However, imagining the existence of nothing in the backdrop of (Universal) Nothing (analogously, finding a black snake in a dark environment) or allowing the mind to remove everything including even one's own body—one thing after the other by the process of successive exclusions (or, simply allowing things to vanish all at a time)—could *be much tougher for most of us*, the human beings—primitive, historic, and modern. This needs an extraordinary sense of *detachment* (meaning giving up the notion of "I" and "mine" referring not so much to the renunciation of possession but renouncing the idea of possessor) and *spirituality*.

1.4 Zero is the very life of all sciences and engineering

Zero is very much more extensively known than the famous constants such as Pi, e (exponential function of argument 1), and Phi (Golden ratio). Everything in any science, any engineering, and any technology will simply collapse and die readily if zero is taken out (unlike the numbers Pi, e, and Phi). Even an irrational number (having infinity of digits), such as Pi, e, and Phi, which contain zeros in their numerical values, will become nonrepresentable (as a number) if zeros are dropped. Not only in the conventional decimal number system, but also *in any other number system of any radix, the symbol of zero along with its unique physical meaning is preserved.* This is not so true with any other symbol implying a nonzero number, say, 11 (in octal, i.e., in base-8 (i.e., radix-8) number system, its physical meaning is 9 and in binary, i.e., in base-2 number system, its physical meaning is different and it is 3), while 00 in any number system of any positive integral radix (e.g., 8, 2, 16, 20, 60) has its physical meaning preserved, that is, it is always 0.

1.5 Nomenclature, symbols, and terms concerning zero and place–value system

The word *zero* came from Venetian *zero* via French *zero*, which (together with cipher or, equivalently, cypher) came, via Italian *zefiro* from Arabic *safira* meaning "it was empty" or, equivalently, *sifr* (the Persian mathematician Mohammed ibn-Musa *al-Khwarizmi* (around 780–850 AD) called zero "sifr," from which our cipher is derived.) denoting "zero" or "nothing." This was a translation of the Sanskrit word *śūnya* (*shoonya* meaning "empty"). *Brahmagupta* (born 30 BC), a renowned Indian mathematician and astronomer and author of many important works on mathematics and astronomy, used *dots* or, equivalently, points (a dot is called *bindu* in Sanskrit and many other Indian languages such as the Bengali language) underneath numbers to indicate a zero. These dots were alternately referred to as "*sunya*." which means empty, or "kha," which means place. Much earlier (more than 2700 years earlier than Brahmagupta) *Aryabhatta*, (born 2765 BC in Patliputra in Magadha, modern Patna in Bihar), the Indian mathematician and astronomer, taught astronomy and mathematics when he was 23 years of age, in 2742 BC. He devised a number system which has no zero yet was a positional system. He used the *word "kha" for position and it would be used later as the name for zero*. There is evidence that a dot had been used in earlier Indian manuscripts to denote an empty place in positional notation. It is interesting that the same documents sometimes also used a dot to denote an unknown where we might use *x*. Later Indian mathematicians had names for zero in positional numbers yet had no symbol for it. The first record of the Indian use of zero which is dated and agreed by all to be genuine was written in 876 AD. This does not imply that before and even a long time before 876 AD the Indian use of zero did not exist. *Aryabhatta* stated that "*sthānāt sthānaṁ* daśaguṇaṁ *syāt*," that is, "from place to place each is ten times the preceding," which is the origin of the *modern decimal-based place value notation*. He devised a positional number system in which the word "kha" was used for position and later as the name for zero. Thus he made use of decimals, the zero (sunya), and the place–value system. Hence the concept of zero as we know today was very much there during his time (viz. third millennium BC). *Bhaskara I* (before 123 BC) is the earliest known commentator of Aryabhatta's works. His exact time is not known, except that he was in between Aryabhatta and *Varahamihira* (Varahamihira, working 123 BC, was born in Kapitthaka or Ujjain, India, and was a Maga Brahmin.

He was an astronomer, mathematician, and astrologer. His picture may be found in the Indian Parliament along with Aryabhata. He was one of the nine jewels (Navaratnas) of the court of legendary King Vikramaditya (102 BC–18 AD)). Bhaskara I mentions the names of *Latadeva*, *Nisanku*, and *Panduranga Swami* as disciples of Aryabhatta. Moreover, he says that Aryabhatta's fame has crossed the bounds of the oceans and his works have led to accurate results, even after so much time. This shows that Bhaskara I lived much later than Aryabhatta. His works are *Mahabhaskariya*, *Aryabhatteeyabhashya*, and *Laghubhaskariya*. He (Bhaskara I) was the first to write numbers in the Hindu–Arabic decimal system with a *circle for the zero during the second century BC*.

Fibonacci (Leonardo of Pisa, or "son of Bonacci," around 1170–1250 AD, was born in Pisa, Italy, and was reared as a teenager in Bougie on the Algerian coast in North Africa, where his father was a customs agent for Pisan merchants) is credited with introducing the decimal system to Europe, and used the term *zephyrum*. This became *zefiro* in Italian, which was contracted to *zero* in Venetian.

About the same time (thirteenth century AD) Jordanus Nemerarius was introducing the Arabic system into Germany. He kept the Arabic word, changing it slightly to *cifra*. The attitude of the common people toward this new numeration is reflected in the fact that soon after its introduction into Europe, the word cifra was used as a secret sign; but this connotation was altogether lost in the succeeding centuries. The verb decipher remains as a monument of these early days.

As the decimal zero and its new mathematics spread from the Arabic world to Europe in the Middle Ages (Medieval Period, fifth–fifteenth century AD), words derived from *sifr* and *zephyrus* (when the concept of zero arrived in Europe, the Arabic word was assimilated to a near–homophone in Latin, zephyrus, meaning "the west wind" and, by rather convenient extension, a mere breath of wind, a light breeze, or—almost—nothing) came to refer to calculation, as well as to privileged knowledge and secret codes.

According to *Georges Ifrah* (Georges Ifrah, born 1947, in Marrakech, French Morocco is a French author and historian of mathematics, especially that of numerals. He was formerly a teacher of mathematics. His exhaustive work, *From One to Zero: A Universal History of Numbers* (1985, 1994) was translated into several languages, became an international bestseller, was included in *American Scientist*'s list of "100 or so Books that shaped a Century of Science," referring to the twentieth century. Despite popular acclaim for his works on the history of numbers, they have been criticized by a few scholars), "in thirteenth-century Paris, a 'worthless fellow' was called a '… cifre en algorisme,' i.e., an 'arithmetical nothing'." From *sifr* also came French *chiffre* = "digit," "figure," "number," *chiffrer* = "to calculate or compute," *chiffré* = "encrypted." Today, the word in Arabic is still *sifr*, and cognates of *sifr* are common in the languages of Europe and southwest Asia. The first known English use was in 1598.

There are different words used in modern times for the number or concept of zero depending on the context. For the simple notion of lacking, the words *nothing* and *none* are often used, while *nought*, *naught*, *aught*, and *ought* remain in use for zero in specific situations. The word *aught* continues in use for 0 in a series of one or more for sizes larger than 1. For American Wire Gauge, the largest gauges are written 1/0, 2/0, 3/0, and 4/0 and pronounced "one aught," "two aught," "three aught," and "four aught." Shot pellet diameters 0, 00, and 000 are pronounced "aught," "double aught," and "triple aught." Decade names with a leading zero (e.g., 2000 to 2009) were pronounced as "aught" or "nought." This leads to the year 2005 ('05) being spoken as "[twenty] aught five" or "[twenty] nought five." Another acceptable pronunciation includes "[twenty] *oh* five." There are archaic and poetic forms with the same meaning. Several sports have specific words for zero, such as *nil* in soccer and football, *love* in tennis and badminton, and *a duck* in cricket. In British English, it is often called *oh* in the context of a telephone number and a point in time (e.g., eight oh five, i.e., 8:05).

Slang words for zero (nothing) include *zip* (received zip for money after doing the job for Jacob), *zilch* (Sam knows zilch about art), *nada* (nothing in Portuguese, Galician, and Spanish), *scratch* (from the very beginning), and even *duck egg* (an opening batsman ignominiously dismissed for a duck egg) or *goose egg*.

The modern numerical digit 0 is usually written as a *circle* or egg-shaped symbol/*ellipse*. Traditionally, many print typefaces made the capital letter O more rounded than the narrower, elliptical digit 0. Typewriters (now almost obsolete) originally made no distinction in shape between O and 0; some models did not even have a separate key for the digit 0. The distinction came into prominence on modern character displays.

A *slashed zero* (usually nonexistent on a modern computer keyboard but sometimes used in hand-written notes) can be used to distinguish the number from the letter. The digit *0 with a dot in the center* seems to have originated as an option on IBM 3270 (a class of block oriented computer terminals, sometimes called display devices, made by IBM originally introduced in 1971 AD and normally used to communicate with IBM mainframe computers) displays and has continued with some modern computer typefaces such as Andalé Mono, and in some airline reservation systems. One variation uses a *short vertical bar* instead of the dot. Some fonts designed for use with computers made one of the capital-O–digit-0 pair more rounded and the other more angular (closer to a rectangle). A further distinction is made in falsification-hindering typeface as used on German car number plates by slitting open the digit 0 on the upper right side. Sometimes the digit 0 is used either exclusively, or not at all, to avoid confusion altogether.

As a cardinal number, that is, a number indicating quantity but not order in a group (e.g., six as distinct from sixth), zero, as stated above, has been denoted by various symbols/words such as 0, zero, Oh, nought, naught, nil, and love (score of nothing as in tennis, squash, and badminton). As an ordinal number, that is, a number showing a position in a sequence/series, zero has been expressed as terms such as *0th*, *zeroth*, and *noughth*.

Different languages such as Arabic, Asamese, Bengali, Devanagari (Sanskrit), Gujarati, Gurumukhi, Kannada, Lepcha, Malayalam, Nepali, Oriya, Tamil, Telugu, Chinese, Japanese, and Thai have different symbols for zero, but the majority of languages seem to use a round circle or an oval-shaped symbol or a variation of them to denote a zero. In Khmer, a small circle (○) and in Thai, a relatively big circle (◯) are used for zero, whilst in Urdu, a solid diamond (◆) is used for zero. In number systems such as binary, octal, decimal, duodecimal (Base 12 or dozenal), and hexadecimal (Base 16), the symbol 0 as available on an English computer keyboard for zero is used and as a single digit it has identical meaning in all number systems with integral radix ≥ 2.

The Bakhshali Manuscript (about 200 BC) was found in 1881 in the village Bakhshali in Gandhara, near Peshawar, North–West India (present–day Pakistan). This document contains many examples of numbers written using the sign zero and the place–value system, as well as several numerical entries expressed in numerical symbols.

In the *Puranas*, great importance is placed in decimal numeration. Thus, in *Agnipurana*, the eighth text, during an explanation of the place–value system, it is written that "after the place of the units, the value of each place (sthana) is ten times that of the preceding place." Similarly, in the *Shivapurana*, it is explained that usually "there are eighteen positions (sthana) for calculation," the text also pointing out that "the Sages say that in this way, the number of places can also be equal to hundreds." These cosmological–legendary texts have often been dated from the fourth century BCE (Before the Common Era, also known as BC or B.C. which stands "Before Christ"), and some have even been dated as far back as 2000 years BCE. These dates, however, are unrealistic, because these texts are from diverse sources and they are the fruit of constant reworking carried out within an interval of time oscillating between the sixth and the twelfth centuries CE (Common Era, also known as AD or A.D. which stands for Anno Domini meaning the "year of our Lord").

The terms *Shunyata* and *Sthanakramad* have been in use in Sanskrit and in many Indian languages at least since the Vedic era.

1.5.1 Shunyata

In Sanskrit, the privileged term for the designation of zero is shunya, which literally means "void." But this word existed long before the discovery of the place–value system. Since antiquity, this word has constituted the central element of a mystical and religious philosophy, developed as a way of thinking and behaving, namely the philosophy of "vacuity" or shunyata. This doctrine is a fundamental concept of Buddhist philosophy and is called the "Middle Way" (Madhyamaka), which teaches that everything in the world (samskrita) is empty (shunya), impermanent (anitya), impersonal (anatman), painful (dukha), and without original nature. Thus this vision, which does not distinguish between the reality and nonreality of things, reduces these things to complete insubstantiality.

This philosophy is summed up in the following answer that the Buddha is said to have given to his disciple Shariputra, who wrongly identified the void (shunya) with form (rupa): "That is not right," said the Buddha, "in the shunya there is no form, no sensation, there are no ideas, no volitions, and no consciousness. In the shunya, there are no eyes, no ears, no nose, no tongue, no body, no mind. In the shunya, there is no colour, no noise, no smell, no taste, no contact and no elements. In the shunya, there is no ignorance, no knowledge, or even the end of ignorance. In the shunya, there is no aging or death. In the shunya, there is no knowledge, or even the acquisition of knowledge."

1.5.2 Sthanakramad

Sthanakramad is a Sanskrit term which literally means "in the order of the position." Often used by Indian scholars in ancient times (fifth–seventh century CE) to indicate that a series of numbers or numerical word–symbols were written according to the place–value system. An example of this is found in the Jaina cosmological text, the Lokavibhaga ("Parts of the Universe"), which is the oldest known Indian text to contain an example of the place–value system written in numerical symbols.

In the fifth century CE, the first nine Indian numerals taken from the Brahmi notation began to be used with the place–value system and were completed by a sign in the form of a little circle or dot which constituted zero: this system was to be the ancestor of our modern written numeration.

1.6 Special terms concerning zero/infinity

There are several special terms that are in use in specific contexts. Some of them are as follows.

1.6.1 Zero for blast

The space shuttle always waits for zero before it blasts into the air.

1.6.2 Ground zero

When we drive toward the site where a bomb went off, we are approaching ground zero.

1.6.3 Zero hour

While the term zero hour has been used in different contexts, it refers to Midnight, or 00:00 hour, in general. The term is implicit in the 2012 American academy award winning (at the 85th Academy awards) action thriller war film *Zero Dark Thirty* directed by Kathryn Bigelow and written by Mark Boal. Billed as "the story of history's greatest manhunt for the world's most dangerous man," the film dramatizes the decade-long manhunt for a terrorist leader after the September 11, 2001 terrorist attacks in the United States. This search eventually leads to the discovery of his hideout and the military raid on it that resulted in his death. The film's working title was *For God and Country*. The title *Zero Dark Thirty* was officially confirmed at the end of the film's teaser trailer. Bigelow has explained that "it's a military term for 30 minutes after midnight, and it refers also to the darkness and secrecy that cloaked the entire decade-long mission." However, the film has received criticism for historical inaccuracy. Former Assistant Secretary of Defense Graham T. Allison has opined that the film is inaccurate in three important regards: the overstatement of the positive role of enhanced interrogation methods, the understatement of the role of the Obama administration, and the portrayal of the efforts as being driven by one agent battling against the CIA "system."

1.7 Digital display A 7-segment display

Figure 1.1 shows a form of an electronic display device to display all decimal digits including zero (and 26 alphabetic characters A, B,..., Z), that is an alternative to a more complex dot matrix display. Seven-segment displays (7SDs) are widely employed in many digital devices, such as digital clocks/watches, electronic meters (e.g., speedometers), electronic pocket calculators, and many other electronic devices, such as household appliances, that display numerical information.

On a 7SD of a calculator, a digital watch/clock, a digital speedometer, or a household appliance, zero is usually written with six line segments, on some older models, it was written with 4 line segments though. The Chinese and Japanese symbols for zero are 零 and 零.

English is no longer the language of Englishmen/British people only, it has become the de facto international language (unlike any other language) which is used by most people practically in all kinds of communications/areas, such as those in all sciences and engineering research, finance, banking, and inter-people. It is difficult to imagine a renowned scientist who is not knowledgeable in English. However, in most natural languages of the world people usually use the numeric symbols 0, 1, 2, 3, 4, 5, 6, 7, 8, and 9 instead of their own language numeric symbols. For example, there are 179 languages in the Indian subcontinent. Most of these languages use the foregoing symbols 0, 1,..., 9 instead of their own language numeric symbols, in general. Such a practice of using only one kind of numeric symbols (without mixing with the regional symbols) is desirable to avoid confusion. Thus the symbol 0 for zero is used more or less universally in most well-known languages of the globe.

1.8 Division by exact zero and nonexact zero

The divisors of zero (*meaning exact/absolute zero unless otherwise stated*) are all numbers except itself. On the other hand, division of any number (negative, zero, or positive) by zero is prohibited. The very first and the most important commandment

Figure 1.1 7-segment display of 10 decimal digits including zero.

(law) in mathematics is "*Thou shalt not divide by zero.*" In fact, in a physical world, division by zero, that is, absolute zero is mathematically illegal and amounts to violation of *a law of nature* (since, if permitted, no meaningful or unique physical interpretation that fits in well with other nonzero numbers in all situations, e.g., in elementary algebra, is possible). *In natural mathematics (mathematics done by nature), division by zero will never occur since nature never violates any of her laws under any circumstances and at any time.* If the concerned zero is a local zero, that is, a relative or nonexact zero, or, equivalently, a *numerical zero* (and not the absolute zero), then one may attach a physical interpretation of the division by such a zero. Such an interpretation is, however, context (including the specified precision of the computer) dependent, where the numerator could be the (absolute) zero or a higher-order (sufficiently smaller) zero (e.g., zeros occurring in the division in an iteration of Newton's second order fixed-point iterative scheme for a multiple root of a polynomial equation on a computer).

1.8.1 z/0 = 0 for any z?

Kuroda et al. (2014) provided new interpretations of the division by exact zero. According to these interpretations, $z/0 = 0$ for any z (including $z = 0$) in their paper "M. Kuroda, H. Michiwaki, S. Saitoh, and M. Yamane, New meanings of the division by zero and interpretations on $100/0= 0$ and on $0/0 = 0$, *International J. Applied Mathematics*, *27*, No. 2, 2014, 191–198." Not only a mathematician but also a physicist (or, for that matter, any scientist/engineer) would immediately react in a violent way since the meanings shake any scientist's extremely deep-rooted conviction, viz. "Thou shalt not divide by zero," that existed throughout his life. After reading their paper, the scientist might ponder over the new explanations provided by the authors using examples taken from physics and also from mathematics. Consider, for instance, the following law of physics. If m_1 and m_2 are two masses situated at a distance r, and if c is a constant, then the (positive) force of attraction between the two masses is $F = cm_1m_2/r^2$. When r tends to $0+$ (in the limit), $F = \infty$ (infinity). However, the authors interpret the force $F = cm_1m_2/0 = 0$. When this law of gravitation was discovered, r was implicitly assumed to be not equal to zero. For if it is zero, then really we do not have two distinct masses m_1 and m_2. We simply have one body with one mass (m_1+m_2, probably). Consequently there is no question of having a force of attraction (assuming the existence of only two masses in the universe). As a matter of fact, the law of gravitation simply remains nonexistent (when there is only one body). In Nature, we will never have any division by exact zero; at least we have not come across any situation in Nature where division of a nonzero (or 0) quantity by exact zero occurs. Also, all the laws of Nature are perfectly followed by any event/activity (including hurricanes, earthquakes, nuclear blast, and death of a star/planet) all the time eternally without absolutely any violation. Strictly speaking, if any mathematical model (which often is designed and developed using no assumption (an assumption always distorts the actual physical problem)) represents the physical problem (occurring in Nature) exactly, then division by exact zero must not be occurring in the model. Hence the question of division by exact zero should never occur in a mathematical model. If it occurs at all, then one should fall back on the model and find out the reason(s) why it has happened and obviate this situation (division by zero) and replace the old (unacceptable) model by the revised one (that does not involve division by zero). If we, on the other hand, still proceed allowing the division by the exact zero with the result 0 then it could be a computational disaster in any digital (ever finite precision) computer. When Matlab encounters such a situation due to a faulty model or a programming mistake, then it comes out with the error "*NaN* (Not a Number)" to warn the user that he has committed a mistake/blunder in the mathematical model (or/and the concerned computer program) that represents a problem in Nature. Thus the commandment in mathematics (conventional/natural/computational), viz. "*Thou shalt not divide by zero*," remains valid eternally. Although we do appreciate the arguments/view points of the authors, these will not affect/alter the present state of mathematics/computations.

In Chapter 2, we demonstrate why zero is a landmark discovery unlike other nonzero numbers such as 1, 5, and 9 on one hand, and why it could conjure up the extremely fearful void and also represent the highest state of mind, viz. the state of silence, on the other. In Chapter 3, we describe representation and a brief history of zero including its role as a number, a place holder, and an operand in arithmetic operations, besides landmark innovations during four distinct periods since 7000 BC. Various uses of zero in sciences, engineering, and in different countries with a stress on Indian zero, besides significant opposition faced by it (zero) are recorded in this chapter. Included in Chapter 4 are the numerical and natural mathematics in the background of conventional mathematics related to zero with a discussion on *calculus* as an ultimate step in understanding the zero in mathematics. Added further in this chapter are a zero in (numerical) computational mathematics and that in mathematics along with their role. This is because these are important over centuries and more specifically in modern times with the advent of the all-pervasive hyper-speed digital computers. These computers are inseparably connected with all walks of human lives and our civilization. This is unlike the period starting from the beginning of the prehistoric age that includes stone age, bronze age, and iron age till late twentieth century. Chapter 5 comprises conclusions.

Zero a landmark discovery, the dreadful void, and the ultimate mind: Why

2

We first observe that *zero and the wheel* are the two most outstanding discoveries that lifted the human beings from other creations to a much greater height in the history of civilization. We discuss below this aspect along with the reasons why zero is a landmark innovation in our mundane aspect on one hand, and the dreadful nonexistence and the highest state of mind in our spiritual world on the other.

2.1 A landmark discovery

Zero and the wheel were the first two *greatest discoveries* in the history of mankind. The wheel brought about the revolution in transportation and machines while zero introduced a revolution in mathematics and computations. Both changed the life of mankind. Both accelerated the human civilization and brought it to a much greater height. Both zero and the wheel coincidentally look "circular."

The status of zero and that of numbers such as 1, 2, and 3 are not the same. For a long time starting from the prehistoric days, people had no problem grasping the concept of one or two or three things and some kind of easily understandable representation of them. While "nothing" such as no cows is not difficult for people to fully realize, its representation and its compatibility (with other nonzero numbers) as a number, a symbol, a direction separator, a magnitude, and a place holder, five-in-one operating with a fully established positional number system (as we are accustomed today), were not readily realizable for a long time as per the available records.

The above-mentioned five attributes happen to be all that a zero should have for its integration with all other numbers in terms of its perfect blend with any context anywhere in science and engineering. No other attribute appears to be necessary to be included for zero for compatibility with other numbers, nonnumerical text, and for its representation. To identify these five distinct attributes by mathematicians and others took centuries, although they came very close to achieve this breakthrough and then deviated away from it.

The foregoing five attributes are compatible with the following notions.

1. Zero is the integer immediately preceding 1.
2. Zero is an even number because it is divisible by 2.
3. Zero is neither positive nor negative.
4. Zero is a natural number (by most definitions), and then the only natural number not to be positive.

Zero: A landmark discovery, the dreadful void, and the ultimate mind. DOI: http://dx.doi.org/10.1016/B978-0-08-100774-7.00002-8

5. Zero is a number which quantifies a count or an amount of null size.
6. Zero was identified before the idea of negative things (quantities) that go lower than zero was accepted (in most civilizations).
7. The *value zero is not the same as the digit zero* when used in a positional number system. Successive positions of digits (reading from right to left as in English and many other languages in the world) have higher weights, thus inside a k-digit number the digit zero is used to skip a position and give appropriate weights to the preceding and following digits. A leading zero digit, in, say, two-digit number 05, is not always necessary in the system. A leading zero may, however, be used for convenience in a given context.

2.2 The dreadful void!

Was there any point in time when it was a complete void—no land, no ocean, no air, and no sky? Let us just think about nothing or emptiness. Everything material—the moon, the sun, the earth, and the whole universe—is vanishing to nothingness. The land under our feet is becoming nonexistent. Even the bodies that I and you have are vanishing. Imagining a situation like this and believing completely this state could terrify one unless he has the requisite spiritual attainment/strength and/or detachment. The following incident in the life of *Swami Vivekananda* (SV, 1863–1902 AD), whose premonastic name was Narendra Nath Datta (Naren) and who is one of the most well-known modern saints/spiritual scientists of India, is significant.

2.2.1 *True incident in the life of Swami Vivekananda in 1881*

For quite a long time, Sri Ramakrishna (SR, 1836–1886 AD), mystic saint as well as the teacher par excellence of Naren, was eagerly expecting the arrival of his disciples, and at the very first meeting with Naren at Dakshineswar, Kolkata in November 1881 AD, he readily recognized in him the worthiest of them all. The second time Naren—a 19-year-old man with an extraordinarily strong mind and an outstandingly sharp intellect—went to Dakshineswar, a month later, SR, the Master, was alone, sitting on his bedstead. As soon as he saw Naren, he received him cordially and asked him to sit near himself on the bed.

In a moment, overcome with emotion, the Master drew closer to him. Muttering something to himself, and with eyes fixed on the young aspirant, he touched him with his right foot. The magic touch produced a strange experience in Naren. *With his eyes open, he saw the walls and everything in the room, nay, the whole universe and himself within it, whirling and vanishing into an all-encompassing void. He was frightened as he thought he might be on the verge of death, and cried out: "What are you doing to me? I have my parents at home."* SR laughed aloud at this, and stroking Naren's chest, said: "All right, let us leave it there for the present. Everything will come in time." Surprisingly, as soon as he uttered these words, Naren became his old self again. SR, too, was quite normal in his behavior towards him after the incident, and treated him kindly and with great affection.

To conceive of this universal nothingness has been attempted possibly by many at different times in modern, historic, and prehistoric ages using the increasingly calmer mind.

Something existing in the background of nothingness (or the other way) is probably easier realizable as there is a contrast. If there is no contrast, that is, nothing existing in the backdrop of an infinite void, then conceiving such a state is truly terrifying (as even one's own body is whirling and disappearing into the limitless void)!

2.3 The ultimate mind

The reader would appreciate better keeping in mind the clear background of the incident of Swami Vivekananda, a most well-known spiritual giant of modern India, once again. His attaining Nirvikalpa Samadhi (NS)—a no-thought state—at the age of 22 in Kolkata, India and his early life are described in brief to understand his extraordinary control over his mind and his outstanding personality. An attempt is made to define the state of NS in the spiritual plane as the equivalent of the Bose–Einstein condensate of the neuronal system in the physical plane.

2.3.1 Nirvikalpa Samadhi and Bose–Einstein condensate

Spiritual and materials sciences are closely connected in the sense that spiritual transformation gets reflected in the very physiological functioning of the body system of any living being. The effect of NS which SV (Swami Vivekananda or simply Swamiji) attained during the last days in the life of Sri Ramakrishna in/around 1885 in Kolkata, India can be measured scientifically only through the state of matter constituting the concerned body. The activity or, equivalently, the kinetic energy of the atomic/subatomic particles constituting the biological living body of a yogi/yogini tends to get converted to potential energy as he/she marches toward silence (NS) or no-thought state.

We stress the fact that there is a Himalayan difference between a deep sleep, presumably a no-thought state, and the state of NS. The latter endows the living being with the highest perpetual wisdom, when he/she returns to a state of nonsilence. We have attempted to look deeply into the effect of concentration on our nervous system, culminating finally into NS. It is interesting to view the effect of NS in the spiritual plane in terms of the Bose–Einstein condensate, that is, the zero kinetic energy state in the physical plane.

2.3.2 Swami Vivekananda in the making

Swami Vivekananda (1863–1902), an outstanding spiritual scientist of modern India, was born Narendra Nath Datta, or simply Narendra or Naren as he was known during his premonastic days, on January 12, 1863 in Kolkata, Bengal (now West Bengal), British India into an aristocratic Bengali family. His father Biswa Nath Datta was a well-known attorney-at-law in the Kolkata High Court and his mother Bhubaneswari Devi was a very intelligent and pious lady, an accomplished woman with a regal bearing and endowed with a keen memory. The Datta family was rich, respectable, and renowned for charity, learning, and a strong spirit of independence.

Naren's grandfather, Durga Charan Datta, was proficient in Persian and Sanskrit as well as in law. But after the birth of his son Biswa Nath, he renounced the world and became a monk at the age of 25. Biswa Nath was well-versed in English and Persian, and took great delight in reciting to his family the poems of the Persian poet Hafiz-e Shirazi. He also enjoyed the study of the Bible which he thought contained the highest wisdom. He was charitable, sometimes extravagantly, and sympathetic toward the poor and those who suffered.

Naughty and restless though Naren was by nature, and given to much fun and frolic, he was greatly attracted toward spirituality even in childhood.

In 1871, at the age of eight, he joined the ninth class at Metropolitan Institution of Ishwar Chandra Vidyasagar (1820–1891), a renowned social reformer of nineteenth century British India. Naren was gifted with multiple talents and he cultivated them all. "His leonine beauty was matched by his courage. He had a delightful voice, the built of an athlete, and a brilliant intellect." His interests included cooking, fencing, wrestling, rowing, games, physical exercise, organizing dramas to instrumental and vocal music, and the love of philosophic discussion and criticism. In all these he excelled to such an extent that he was an undisputed leader. These and other traits in his character soon attracted the notice of his teachers, and fellow students/friends. At college Naren took studies beyond the college syllabi more seriously. He was a voracious reader with an extraordinary concentration and exceptional critical analyzing power coupled with an unusual capacity of assimilating the authors' thought process exceptionally fast. He keenly studied the works of David Hume, Johann Gottlieb Fichte, Baruch Spinoza, Georg W. F. Hegel, the positivist philosophy of Auguste Comte, western logic, the abstruse philosophy of Herbert Spencer, the systems of Immanuel Kant and Arthur Schopenhauer, the mystical and analytical speculations of the Aristotelian school, and John Stuart Mill's *Three Essays on Religion*. He also mastered the English poets like Percy Bysshe Shelley and William Wordsworth and the ancient and modern history of Europe.

He even took a course in physiology with a view to understanding the functioning of the nervous system, the spinal cord, and the brain. Besides studying western philosophers, he was thoroughly acquainted with Sanskrit scriptures and many works in Bengali.

But this exposure to western philosophy which lays special stress on the supremacy of reason, brought about a severe conflict in Naren. On one hand, his inborn tendency toward spirituality and his regards for the ancient traditions and beliefs which he had embodied from his mother, and on the other his argumentative nature coupled with his sharp intellect and deep insight, which abhorred all kinds of superstition and questioned faith, were now at war with each other. Under a deep spiritual urge, he was then found observing hard ascetic practices, staying in his grandmother's house, away from his parents and other relatives, following a strict vegetarian diet, sleeping on the bare ground or on an ordinary mat, in accordance with the strict rules of *brahmacharya* characterized by the practice of strict celibacy.

Two visions of life had presented themselves before young Naren. In one, he found himself among the great ones of the earth, possessing riches, power, honor, and glory, and he felt himself capable of attaining all these. In the other, he saw himself renouncing all worldly things, dressed in a simple loin-cloth, living on alms, sleeping under a tree, and then he felt that he had the capacity to live this like the Rishis (spiritual scientists) of ancient India. It was, however, the second vision that prevailed in the end, and he used to sleep with the conviction that by renunciation alone could man attain the highest bliss.

He also used to meditate for long hours before going to sleep; and from boyhood he had a passion for purity, which his mother made him observe as a matter of honor, and in loyalty to herself and the family tradition. He was a born idealist and seeker of truth; so though he could hardly be satisfied with worldly enjoyment, he was a jubilant lover of life.

In 1879, he entered the prestigious Presidency College (now Presidency University), Kolkata after passing the Entrance Examination. A year later he joined the General Assembly's Institution (now Scottish Church College) and passed FA (First Art) and BA` Examinations. The principal of his college, Professor William Hastie, once remarked: "Narendra is a real genius. I have travelled far and wide, but have not yet come across a lad of his talents and possibilities even among the philosophical students in the German universities. He is bound to make his mark in life." He was highly impressed by Naren's philosophical insight. It was from Hastie that he first heard of Sri Ramakrishna.

One day in November 1881, Naren went to Dakshineswar, Kolkata to meet Sri Ramakrishna. At the very first meeting with Naren, Sri Ramakrishna immediately recognized in him the one whom he had been waiting for so long.

In 1884, Bishwa Nath Datta suddenly passed away due to a massive heart attack, plunging the whole family into grief and utter poverty. He was the only earning member of the family and being of a prodigal nature, he spent lavishly and left the family in debt. At the time Naren was studying for the BA, and had just finished the examination. As the eldest surviving son, he now had to shoulder the entire responsibility of the family. Starving and barefoot, he went from office to office in the scorching sun in search of a job. Everywhere the door was slammed on his face. Friends turned into enemies in an instant. Creditors began knocking at the door. Often he went without food so that the others at home might have a better share. He was face to face with realities, and the world appeared to him to be the creation of a devil.

2.3.3 Swami Vivekananda attaining NS

It may not be out of place to mention that at an elementary state of mind, thoughts are random or chaotic. Just at this moment, one may have the thought of the basketball match being played at the indoor stadium, while at the next moment the mind may enter into the thought of the forthcoming test in Deterministic Operations Research, and at the third moment the mind may shift to the thought of the poor health of one's mother.

When one reads a book with concentration, there will be a chain of connected thoughts as if the have-has-had (chaotic) orientation of the magnetic molecules (dipoles) of an iron bar depicting no magnetic effect becomes oriented systematically resulting in a powerful magnet. The thoughts are no longer random. This is a higher state of mind which is a more powerful conscious assimilator of information/knowledge than the elementary mind.

If the mind has only one thought, say, the thought of a beautiful scenery or an excellent music, then it is still a higher state well-suited for innovation/revelation when focused on a particular thought/query.

From this state, the mind may drop down to silence (no-thought condition), which is the highest state of mind. There can be nothing higher than this state. At this state, there is nobody to digest the food already existing in one's stomach, nor there is anybody to circulate one's blood. The heart, a small powerful pumping system, which pumps the blood of the whole body four times in a minute throughout the very existence of a human being for over 70 years, say, stops pumping. The difference between the systolic and diastolic pressures vanishes. The lung stops functioning. The body becomes cold like a dead body. The mind at this state has only potential energy, devoid of kinetic energy, but capable of channeling the mental energy to any direction which one desires and achieving the desired result in a much more intense way. This state of silence is called NS.

Only when a thought such as the thought of drinking a glass of water (implanted before the state of Samadhi) enters the mind, does the body start getting heat, and the temperature starts increasing; for instance, heat enters first into the head and then gradually spreads to the lower part of the body, and the mind starts functioning. The potential energy of the mind slowly gets translated to kinetic energy.

Once somebody asked Swami Vivekananda how a no-thought state of mind differs from a deep sleep. Swamiji responded saying that a fool goes into deep sleep and comes out of it as a fool. When the fool enters into the state of Samadhi and then comes out of it, he comes out of it as the wisest person.

To one having experienced this state, revelation comes much more easily when he channelizes/focuses his mind on the concerned issue. The rishis (spiritual scientists) of ancient India and other countries thus have contributed gigantically to the world/society through such revelations. Not only the rishis, but also the materials scientists such as the physicists, chemical scientists, biologists, and medical scientists also had revelations and contributed to the society, consequently transforming the world into the one which we see today. The discovery of the concept of zero and its representation were due to such a revelation that resulted through deep concentration.

The following spiritual events in the life of Swami Vivekananda out of many are relevant as illustrations in this context. The first one occurred in 1885 in Kossipore Garden House in Kolkata and is narrated by Swami Niranjanananda who is one of the sixteen direct disciples of Ramakrishna Paramahamsa (1836–1886), a most renowned nineteenth century mystic.

It was one summer afternoon in Kolkata. Inmates of the garden house were busy. Some were undertaking their personal work while others were preparing for a stroll in the humid hot evening. Narendra Nath Dutta, the premonastic name of Swami Vivekananda, was meditating in the ground floor hall of the garden house lying down with a cloth covering his body. During the meditation, his legs were becoming still and completely numb. His body temperature started receding. Finally his whole body up to the suture on the top of his head became cold. Niranjan Maharaj (Swami Niranjanananda) due to some work went to call Naren and touched him. After close examination, he found Naren's body had become completely motionless and cold like ice as if he had died a long time ago and he had no sense.

Niranjan Maharaj became very apprehensive. He went on running here and there and calling people around. At last he brought a doctor and tried hard to bring back his senses. Everybody was anxious and sad believing that Naren had suddenly left his mortal body. Niranjan Maharaj felt that everybody was running around, but nothing had been told to Paramahamsa who was upstairs in the two storied building. To inform him about the most unexpected demise of Naren was absolutely necessary. Running up to the upper floor Niranjan Maharaj told him (Paramahamsa), "Sir, Naren has died. His dead body has become cold." Without depicting the slightest grief, Paramahamsa (Sri Ramakrishna) started smiling. Niranjan Maharaj was irritated: Naren died but Sri Sri Ramakrishna was smiling!

He returned to the hall downstairs. Then a little bit of heat had accumulated at the suture and slowly the heat started traveling downward to the throat of Naren. Narendra Nath regaining his consciousness saw Ram Chandra Dutta and shouted, "Ramdada, Ramdada, where is my body? where is my body?" Gradually heat flowed into the other parts of his body and Narendra Nath became normal. Everybody was relieved by seeing their beloved Naren as hale and hearty as before. After a little while Narendra Nath went to meet Sri Sri Ramakrishna. When the afternoon incident was narrated to him, Sri Sri Ramakrishna affectionately told Naren, "Hello, you wanted to see Nirvikalpa Samadhi (the ultimate state of mind)! Now you realized! You have a lot to accomplish. Now the key would remain under lock. It would be opened later."

2.3.4 Meerut incident

An incident that occurred in Meerut in the state of Uttar Pradesh in India merits a mention in this context. It was during the late 1880s or early 1890s. Naren was a wandering monk going from place to place mostly on foot in the British Indian subcontinent to gain first-hand experience of the life current, the underlying culture of this vast country, and to feel the pulse of the great old civilization. Gangadhar Maharaj (Swami Akhandananda, 1867–1937), one of the sixteen direct disciples of Ramakrishna Paramahamsa and the third president of the Ramakrishna Mission, a philanthropic and spiritual organization, used to fetch several books from the local library for Naren to read. These were voluminous books. Once the books were read by Naren within a day or two, he used to return these books and fetch a new set of books. This process continued for several days and the librarian had been observing this.

One day the librarian sarcastically asked Gangadhar Maharaj: "What sir, do you borrow these books only to see colourful bindings or to read them?" Gangadhar Maharaj responded saying that he borrows these books for Swamiji (Naren) who reads these books completely. Tauntingly the librarian said, "Oh, surely yes, I have understood!" The words of the librarian hurt the feeling of Gangadhar Maharaj since he loved Naren more than his own life. This sarcastic remark became unbearable for him. He came back and told all that had happened to Naren.

Narendra Nath smiled a little and told, "Call the man; I will show him how to read. See Ganga, some read word by word, others read sentence by sentence.

Do you know how I read? I read paragraph by paragraph." He then asked Gangadhar Maharaj, "You hold this book. I will go on telling what is written in it." Naren went on narrating verbatim all that had been written in the book. Gangadhar Maharaj and the esteemed people of the locality were simply amazed! *Intense concentration and merging oneself completely with the very thought process of the writer allow one to achieve such a feat.*

2.3.5 State of NS and that of zero kinetic energy: equivalence

Albert Einstein (1879–1955) described, nearly one hundred years ago in 1919, the equivalence of mass and energy as "the most important upshot of the special theory of relativity" that lies at the heart of modern physics. According to his famous equation $E = mc^2$, where the constant c is the velocity of light, E and m are the energy and mass of a physical system, respectively. If we choose a unit where $c = 1$, then $E = m$ denoting equivalence of mass and energy. Furthermore, matter can be converted to energy and vice versa. Matter is present in various energy states where the temperature is the function of an energy state. The greater the energy, the higher is the temperature. Water boils when it is heated to higher temperatures. This enhances the entropy or, equivalently, disorder in the water molecules as these are energized. This represents the excitability and chaos of the molecules constituting the matter, viz. water in this case.

Satyendra Nath Bose (1894–1974), a renowned Indian physicist, proposed to Albert if matter was cooled to the lowest temperature, viz. 0 Kelvin (K), that is, the absolute zero which is approximately equal to −273.15 °C or, equivalently, −459.67 °F, then its entropy should decrease to the minimum and matter should reach a zero kinetic energy state. This proposition was later shown to be true. This zero energy state (i.e., absolute zero kinetic energy state) is called the Bose–Einstein condensate. This state of matter is also known as a superatom since the total mass behaves as if it was a single atom. It loses all its characteristics of shape, electric charge, and polarization.

A New Zealand physicist Ernest Rutherford (1871–1937), and a Danish physicist Niels Bohr (1885–1962), developed a way of thinking about the structure of an atom in which an atom looks very much like our solar system. It is known as the Rutherford–Bohr model of Atomic Structure, and was something of a breakthrough in describing the way the atom works. According to this atomic model, an atom consists of positively charged nucleus made of protons and neutrons. Electrons which are equal in number to the protons, revolve around the nucleus in various orbits. This atom is extremely stable due to the electromagnetic force between the positively charged nucleus and the negatively charged electrons. This model was accepted by scientists as a broad model. The hundreds of subatomic elementary particles, such as mesons, leptons, neutrinos, keons, pions, quark, w plus, w minus, and Higgs Boson (initially theorized in 1964 while its existence was tentatively confirmed on March 14, 2013)—some are low mass, some others are medium mass, and the rest are heavy mass—were observed later over the decades.

At a room temperature (usually 24 °C or 75.2 °F or 297.15 K), the electrons revolving in different orbits around the nucleus are at relatively very large distances like our planets orbiting the sun at different orbits. When temperature is lowered, the distance between an electron and the nucleus becomes smaller. Also the kinetic energy of a revolving electron decreases and the volume of the atom reduces. When the temperature is lowered down to 0 K (or rather very close to 0 K as exactly 0 K is not reachable), the whole atom reduces to a mass having a numerical zero volume and a zero energy state. In the scale of Kelvin, there is no negative temperature unlike that of Celsius and Fahrenheit. It may be remarked in this context that while there is the lower bound of temperature, that is, 0 K, the upper bound of temperature beyond which there cannot be any temperature attainable is not yet known.

A system at absolute zero still possesses quantum mechanical zero-point energy, the energy of its ground state. The kinetic energy of the ground state (or the concerned entropy) cannot be removed. However, in the classical interpretation, it is zero and the thermal energy of matter vanishes. Scientists have achieved temperatures extremely close (within millidegrees or possibly nanodegrees) to absolute zero, where matter exhibits quantum effects such as superconductivity and super-fluidity. The matter, when subjected to withdrawal of heat such that the temperature reaches 0 K, possibly returns to a shapeless and an attributeless phenomenon. In a devolution, it reverts just to a potential energy form to manifest as something not exactly experienced by us.

The human brain consists of 10^{11}, that is, 100 billion, neurons (nerve cells). The thoughts that constantly crowd our minds are due to the sum total of activity of different neurons concurrently. These thoughts are chaotic. The thoughts translate into different biological changes catalyzed by the hormonal apparatus at the hypothalamic-pituitary-adrenal axis interfacing system. The huge number of complex interconnections among the neurons in the nervous system ensures that even a tiny impulse quickly spreads.

There are some individuals who have a higher innate entropy level or, equivalently, higher chaos in their thoughts. These individuals find it hard to concentrate compared to those who have a lower innate entropy level. The former individuals are easily distracted. Their mind roams from one thought to another totally unconnected; they are restless. The sensory organs of these individuals excite various natural neural circuits resulting in a higher entropy. These organs serve as a vital pathway to enhance the entropy. Hence shutting down these organs, such as closing the eyes, could help in the process of concentration, the mind is supreme though.

With the increase in concentration, the disorder of the activities of neural circuits decreases; so does the wavering of the mind. The mind is more keenly focused. As stated earlier, if the mind has only one thought, then it is still a higher state well-suited for innovation/revelation. So when we concentrate intensely, we enhance the synchronicity of a certain group of neurons and silence unrelated neural activities. Just as the entropy of matter drops down to a numerical zero when we march downward to absolute zero, the disorder in the activity of the neural system keeps declining as we concentrate more intensely. The neuronal firing diminishes in amplitude and also in frequency. Further the propagation across various neural networks decreases.

Meditation is nothing but conscious concentration in the wakeful state. This is nothing but a conscious attempt to decrease the entropy of the nervous system. As the entropy of the neurons keeps waning, a state of calmness is perceived.

As this proceeds further, the neurons start becoming synchronous. In other words, the neurons neither modulate nor amplify incoming signals. These resonate in harmony. As this orchestra becomes more intense in synchronicity, the nearer we are to the zero entropy. We experience varying states of bliss and happiness. When all the 100 billion neurons are in complete unified quantum coherence, we arrive at the state, presumably the final state, of zero energy. The mind drops down to silence, the highest state of mind attainable by a human being. The zero energy state, that is, the Bose–Einstein condensate-equivalent of the neuronal system is what may be called the state of NS or, equivalently, the ultimate/zero state of mind.

We have presented in brief the early life of Swami Vivekananda including his momentous association with Sri Ramakrishna, his teacher par excellence. This is followed by a short description of his attaining NS at a tender age of 22 and then by an emphasis on the Bose–Einstein condensate-equivalent of the neuronal system which may be termed as the state of NS or the highest state of mind. In this context, a brief note of the following terms, statements, and questions enhance further clarity for still better appreciation of the consequences of NS manifested on the human biological system and its functioning.

2.3.6 Deep sleep

The information processing rate in our wakeful state is around 10 bits (binary digits) per second while it is 66 bits per second in sleep state. This implies that the mind is much more active in a sleep state than in a wakeful state. In deep sleep, it is not that mind completely hibernates without any activity. The only thing is that the person who enters into deep sleep, comes out of it without any memory of all that has happened during the deep sleep. The physiological functions of the body system such as the food digestion and blood circulation do not come to a standstill, this may be less pronounced though. Thus NS differs from deep sleep in its entirety.

2.3.7 Experience: natural versus artificial

Natural experience involves consciousness of a living being and exists in the infinite domain of knowledge. Artificial experience, on the other hand, is a partial simulation of natural experience and always exists in the finite domain of knowledge. Artificial experience is depicted by a machine, for example, a computer through a computer program written by a human being and the finite data (information) stored into it. Artificial intelligence may be considered as an implication of artificial experience just as natural intelligence is that of natural experience.

2.3.8 Experiencing and not just knowing

There is a gigantic difference between experience and just knowledge. Natural experience is exclusive to a living being and not to a nonliving being. The difference between a living being and a nonliving being is beyond any measure. The story attributed to Swamiji is as follows. A locomotive is moving at a speed of 90 miles per hour. An ant crawling on the rail is trying to run fast to save its life. Which is more majestic: the locomotive or the ant? It is the ant, a living being, which is infinitely more majestic than the locomotive, a nonliving being, it could crush the ant in no time though.

We now consider Dr Watson, the IBM computer, which competed with the world's most knowledgeable human beings, the concerned Americans, and defeated them in "Jeopardy" (an American television quiz show created by Merv Griffin) conducted by Alex Trebek. The huge database consisting of more than 10^{16} bits of information (at least 10 times bigger than the information contained in the British Library) of numerous subject areas of the world/universe can be searched by Dr Watson using the Google search engine—an astonishingly fast (polynomial-time) algorithm, considered to be a wonder of the twenty-first century—or some other fast search routine, in seconds.

The knowledge of human beings (in a conscious state), on the other hand, simply does not/cannot match Dr Watson in speed and also in accuracy. For instance, a specific database consisting of medical information including physiological parameters of a patient and possibly past treatment records obtained from the computer chip embedded in his body can be used by a doctor from any part of the world for the patient in question for prescribing the required treatment. Or, a team of doctors physically located at different parts of the world having simultaneous access to all the information about the patient can confer via video conferencing to decide upon the best future course of treatment in an astonishingly short time. Such a modern, rather ultra-modern, procedure can be argued to be the best in the treatment of a patient. Definitely such a development is a boon compared to what we had before the mid-twentieth century.

Nevertheless a live doctor is prone to error, unlike Dr Watson with artificial intelligence, and being capable of depicting natural intelligence is supreme. After all who has created Dr Watson? It is the combined effort of all those living human beings who are responsible for the birth of Dr Watson, the nonliving computer, which accumulates knowledge from numerous sources physically located all over the globe and beyond dynamically every passing second. Thus the living beings will remain ever supreme. It will not be out of place to state the age-old proverb "To err is human." Similarly we can state "Not to err is computer." Here err implies mistake while human and computer imply a living being and a nonliving being, respectively.

In the case where Dr Watson commits an error, it is not the fault of Dr Watson. It is the fault of the concerned human beings who have written the computer program and entered the concerned data. Such a knowledgeable person, viz. Dr Watson, incapable of natural experience and also devoid of requisite consciousness of a living being will never be able to experience NS and become the wisest person of the world. NS is an exclusive possession of only a living being, specifically a human being. Dr Watson can be made to depict, besides search, artificial intelligence which is entirely based on the gigantic database (usually continuously expanding with time, remaining ever finite though) that he has in his artificial brain, that is, executable and nonexecutable computer memories.

Yet he will not be able to transcend into the arena of natural intelligence (of a living human) where highly vibrant consciousness is the basic constituent. We, the human beings, can simulate to an extent the natural intelligence but we will never be able to bring consciousness and hence natural intelligence to Dr Watson until we are capable of creating him using live biological cells having enormous complexity of inter-connections or, in other words until we inject life into Dr Watson making him a living computer like a truly living human being. As of now injecting such a life is impossible. As a matter of fact, it is the realization of spiritual scientists that we, the human beings, are the greatest in the universe. No other living being in any planet of any galaxy can be superior to us. Such a revelation springs from the highest state of mind, that is, state of NS. Just the existing knowledge will not be enough to realize that we, the human beings, are the greatest in the universe.

2.3.9 Who is faster: man (living computer) or computer in reality?

This question was posed to Sen (one of the authors) during late 1970s by a renowned space scientist who was then the director of Indian Institute of Science, Bangalore. During those days, Sen had been writing computer programs for some of the PhD research students of the institute. While writing a program, we would be writing the instructions sequentially and possibly would be thinking sequentially. He replied saying that the computer is much faster than a human being. The scientist responded: No, it is not true. If you are shown a photo of your mother, how much time do you take to recognize her as your mother? Only a fraction of a second, isn't it?

On the other hand, if the photo is shown to a Cray supercomputer, it would take a couple of seconds to come out with the same answer. This response made him ponder over the question. He felt that in the conscious state of mind, we think sequentially while in the subconscious/unconscious state of mind, we would be thinking in parallel with a speed beyond that of any hypercomputer of today and also of tomorrow.

2.3.10 Neuronal system: natural versus artificial

The brain of a living being, considered to be the most important physical location of mind, consists of approximately 100 billion neurons and their highly complex interconnections, which are beyond the comprehension of any physicist. Thus the minimum, that is, the lower bound, of information that a human being can have is around 100 billion bits if one live neuron stores one bit of information, the smallest possible building block of any information as used in silicon technology today. However, in reality there is no known upper bound of information that one brain can hold. The spiritual scientist has the proof that the amount of information in one brain could have no upper bound.

The proof is his experience, considered to be the flawless best proof, better than even a mathematical proof and as good as a computational mathematical proof. The following quote and our concerned statement are noteworthy.

> *God exists since mathematics is consistent, and the devil exists since we cannot prove the consistency.*
>
> *Morris Kline (1908–1992)*

In the same spirit, we write

God exists since computer mathematics is consistent, and the devil does not exist since we can prove the consistency.

The well-known artificial neural system that consists of artificial neural networks, on the contrary, is a very rudimentary computer simulation of the foregoing neuronal system to seek an answer to a question, which may be crudely acceptable or even unacceptable in a context.

2.3.11 God is omnipresent, omnipotent, and omniscient while computer will never be

By the term "God," we imply consciousness. The term omnipresent implies the unlimited nature of God or His ability to be everywhere at all times while the terms omnipotent and omniscient refer to the all-powerful nature of God and the all-knowing nature of God, respectively. There exists no space including objects—living or not—in the universe where God does not exist. Thus even in a dead human body, consciousness does exist as much as in a live human body. While the manifestation of consciousness is practically nonexistent in a dead human or a dead animal body, it is vibrant in any live body. In some live body, the manifestation is more pronounced than in some other live body. Also, in a single living being—human or not—the manifestation could/would be different at different times. Consequently we may say that any dead body has zero manifestation (not exactly zero) of consciousness. In the same way, all material objects—nonbiological ones—may be considered having zero manifestation of consciousness numerically. A live red blood cell also has its own consciousness (soul) whose manifestation is very much above zero.

Omnipotence and omniscience attributed to God have been the realization/experience of the rishis (spiritual scientists)—known and unknown—of the globe or possibly the universe. The computer on the other hand is the innovation of a mankind endowed with an extraordinary manifestation of consciousness. The manifestation is usually much more pronounced in a scientist—spiritual or materials, renowned or not—than that in an ordinary human being. Thus a computer with all its hyper-speed, hyper-storage, hyper-bandwidth, and the depiction of artificial intelligence, artificial consciousness, and artificial experience will never be able to equate itself to a human being, although he is always prone to error unlike a digital/nonliving computer.

2.3.12 Chaos—does it really exist in nature?

We have not come across any event or any natural phenomenon that has not followed all the laws of nature. The hurricane and the earthquake, for instance, occur following all the laws of motion and gravity—known as well as unknown to us. As a matter of fact, we are familiar with only a few of the laws of nature and even then not always exactly. Nor are we exactly conversant with all the forces acting on a body. Consequently almost all our models are inexact and often prone to provide us with distorted results which could be sometimes unacceptable, these give a better insight for future improvement though.

The particles in a gas or a liquid in a container, which are assumed to be elastic bodies, move randomly hitting one another and also hitting/bombarding the wall of the container. As temperature increases, the movement of each particle increases and therefore so does the number of collisions. All these happen without any disorder or, equivalently, chaos. There is always an order which is mostly beyond our comprehension. However, we need to answer questions important to our society while we are neither equipped with all the laws of nature and all the forces involved in the extremely large n-body problem. Nor are we equipped for correctly formulating and then solving the resulting vast computational problem even if we use the fastest available (over 10^{18} floating-point operations per second) computer of today (2015). The theory of chaos developed based on incomplete and often approximate rules seeks an answer to our queries within the limited human and computing resources and knowledge that we have.

2.3.13 How do we know 0 K which is not reachable?

Using only thermodynamic means, it is not possible to arrive at the lowest possible temperature, 0 K, although we could reach within a few millidegrees away from 0 K. Then how do we know that there is 0 K at which the volume of any matter (gas or solid or liquid) becomes zero or rather a numerical zero? In computational mathematics, this can be known using the method of central difference limit based on extrapolation. This method can be iterated for pronounced accuracy or can be used noniteratively (directly) for an acceptable accuracy.

2.3.14 Experience is the proof

In spiritual science, the proof of an event, for example, the event of NS resulting in the highest wisdom a human being can achieve, is experiencing it. There are different kinds of proof, such as mathematical proof, computer mathematical proof, statistical proof, evidential proof in a court of law, probabilistic proof, documentary proof, and experimental proof. A mathematical proof of a statement, for instance, is based on a method of deduction, that of induction, or that of contradiction, or a combination of one or more of these methods with one or more axioms (statement accepted as true without proof) as inputs. All these kinds of proof procedures, except those for computer mathematics, either suffer from a flaw or an inaccuracy or a deficiency/incompleteness or an incapability or fuzziness.

The experiential proof in spiritual science, on the other hand, is free from all the foregoing drawbacks. One who has actually experienced NS knows that he/she has been endowed with the highest wisdom. Such a person, of course, is almost impossible to be found/detected in our society, although there could be a very few in the midst of us, who usually go unnoticed or undiscovered by us.

2.3.15 Mind is the reservoir of endless knowledge

If we assume that one neuron stores only one bit of information then the human brain or mind can store 100 billion bits of information. This is truly huge but finite and is equivalent to one-tenth of the information held by the British library. If a neuron has to hold information, it cannot be less than one bit which may be called the building block of any information and is not divisible, probably equivalent to an elementary subatomic particle (possibly not divisible further) constituting matter. Definitively the number of neurons comprising the brain as per the scientific measurement is finite and is around 10^{14}.

While the minimum amount of information that can be stored in our brain is 10^{14} bits, there is no maximum amount known to a physicist/computer scientist. But the maximum is known to a spiritual scientist and the proof is his experience. This maximum is infinity of bits of information. Observe that a single natural brain is definitely finite in all physical aspects while its information storage capacity is infinite very much unlike the artificial brain, that is, the machine (computer) brain and its storage capacity, both of which are finite and will ever remain finite, although these can increase exponentially with successive innovations subject to a limit set by the laws of physics. Swamiji declared: All knowledge that the world has ever received comes from the mind; the infinite library of the universe is in our own mind.

2.3.16 Numerical zero versus mathematical zero

Numerical zero is a relative term, context-dependent, and is not unique. The mathematical zero is absolute, context-independent, and is unique. Five million dollars: Is it a big amount of money or a small amount of money? It is a numerical zero compared to the United States federal budget while it is a large amount of money compared to the annual salary of an average professor of the United States.

2.3.17 Consciousness: natural versus machine

The basic question, "What is consciousness from a physicist's, philosopher's, cognitivist's, and informatician's point of view?" has been addressed. Instances are: a physicist attempts to tackle the problem by theoretical means of quantum theory, and a philosopher by a debate on qualia (individual instances of subjective conscious experience such as the perceived brilliant color of early morning sky, the pain of a deep wound, and the taste of Indian food). As stated earlier that consciousness, that is, natural consciousness, exists everywhere, we may call this as the absolute operator which exists in both animate and inanimate objects causing projections, preservations, and destructions. Manifestation of consciousness is considered as the very core of the existence of any living being. According to a scientist/mathematician, the ultimate (manifestation of) consciousness can be attained by an individual, for example, a living human being, by applying the principle of detachment; the philosophical tenets have been analyzed mathematically. while many conjectures and conclusions of Vedanta are fully in conformity with recent scientific investigations.

A computational scientist has at his disposal a finite database. The consciousness derived out of this data, that is, his information base, is artificial. This database is expanding every moment and so the artificial consciousness is also expanding. The database, though continuously growing, remains and will ever remain finite unlike the natural database that a living being is endowed with. The innovation based on the artificial/machine consciousness can be achieved by a machine (computer) through a computer program written by a human being. There is a gigantic, rather infinite, difference between this innovation and the natural innovation.

2.3.18 Measuring manifestation of consciousness

While it is beyond our scope of measuring consciousness in the absolute sense, measuring the manifestation of consciousness in a relative sense numerically may not be altogether outside human capability. As of now there is no known measure of the manifestation. All that we can possibly say is: At a point in time or at a period of time a human being depicts more consciousness than that depicted by another man/woman or than that depicted by himself at some other point in time.

While people such as Ramakrishna Paramahamsha, Swami Vivekananda, Ramana Maharshi, and other renowned/unknown spiritual scientists of the world, as well as the physical scientists, have depicted extraordinary/intense consciousness and contributed very exceptionally to the humanity, others, within their capability, also have demonstrated consciousness to a varying degree and contributed to the society not so exceptionally.

As a matter of fact, there are vital problems, such as measuring the average tension level of people physiologically (not statistically by asking questions to a participant) in a country, say, the United States of America, which are still an important open problem. This problem needs to be solved in order to assess to what extent the development and mechanization in a country is responsible for increasing the stress/tension level of a human being and also possibly the so called quality of living. Such a study could provide the necessary corrective measures or, equivalently, a balance between development and stress level.

History of zero including its representation and role

3

One of the most difficult tasks is to write the history of zero chronologically. Many innovations occurred in different places at different points (or the same point) of time, which are not often precisely known and nor are these chronologically developed, that is, an innovator developed his ideas without often the knowledge already existing elsewhere. In fact, unlike most of today's innovations/discoveries based on already existing contributions, the ancient innovator had at his disposal very little knowledge of all that had existed in a scattered way in different parts of the globe. However, we have categorized four distinct periods during 7000 BC–2015 AD based on most notable innovations—some are overlapped or duplicated—during each of these periods.

3.1 7000–2000 BC: innovation of decimal number system that is universally used today

Probably the most fundamental contribution of *ancient India* to the progress of civilization is the decimal number system including the invention of the number zero. This system uses 10 symbols, that is, nine digits and a symbol for zero, to denote all integral numbers, by assigning a place value to the digits. This system was used in *Vedas* (Vedas are believed to be one of the oldest books (probably developed during 7000–4000 BC) ever made by mankind. They are written in an old Indian language called *Sanskrit*. Vedas and *Puranas* are a vast sea of information and knowledge. Who wrote them and exactly when is difficult to establish but the information contained therein is an embodiment of immense wisdom and amazing accuracy) and in Valmiki *Ramayana*. *Mohanjodaro* and *Harappa* civilizations (3000 BC) also used this system.

The Vedas (*véda in Sanskrit*, meaning "knowledge") are a large body of texts originating in ancient India. Nobody precisely knows the period when Vedas originated and flourished. Composed in Vedic Sanskrit, the texts constitute the oldest layer of Sanskrit literature and the oldest scriptures of Hinduism. The Vedas are *apauruseya* ("not of human agency"). They are supposed to have been directly revealed, and thus are called *śruti* ("what is heard"), distinguishing them from other religious texts, which are called *smrti* ("what is remembered").

If zero denotes a *direction separator*, that is, if it separates those above it from those below it or specifies a magnitude then the *Egyptian* zero, *nfr*, introduced over

Zero: A landmark discovery, the dreadful void, and the ultimate mind. DOI: http://dx.doi.org/10.1016/B978-0-08-100774-7.00003-X

4000 years back served these purposes very well. The ancient Egyptians (5000 BC) did not use positional notation although they used a system based on 10. Thus to represent 428, they would draw four snares, two heel bones, and eight vertical strokes.

The *Sumerians* (Sumer was an ancient civilization and historical region in southern Mesopotamia (*southern Iraq*), during the Chalcolithic and early Bronze age (third millennium BC) were the first to develop a counting system to keep an account of their stock of goods such as cattle, horses, and donkeys, The Sumerian system was positional; that is, the placement of a particular symbol relative to others denoted its value. The Sumerian system was handed down to the Akkadians (the Akkadian Empire was an ancient Semitic empire centered in Akkad and its surrounding region in ancient Mesopotamia) around 2500 BC and then to the Babylonians in 2000 BC. It was the *Babylonians* who first conceived of a mark to signify that a number was absent from a column; just as 0 in 1038 signifies that there are no hundreds in that number. Although zero's Babylonian ancestor was a good start, it would still be centuries before the symbol as we know it appeared.

The renowned mathematicians among the *ancient Greeks*, who learned the fundamentals of their mathematics from the Egyptians, did not have a name for zero, nor did their system feature a placeholder as did the Babylonian. They may have pondered it, but there is no conclusive evidence to say the symbol ever existed in their language. It was the Indians who had understood zero both as a symbol and as an idea.

The representation of zero which, though appears to be a trivial issue today, has been an important requirement to accelerate the speed of the forward movement of all sciences/engineering over which several centuries have been devoted by super-minds in various parts of the world to arrive at its universally acceptable representation. In the history of zero, we limit ourselves mostly in the practical usage aspects of zero here. The history essentially includes that of the number zero (0), its integration with other numbers, and its positional importance when it occurs in a k-digit number.

3.1.1 Prelude

Somewhat unlike the history of very special numbers such as Pi, e (exponential function of argument 1), and Phi (Golden ratio), the history of zero appears to have been much more widely prevalent among not only the serious thinkers including mathematicians/scientists and the mathematical historians but also among very common men who may not be knowledgeable/well-versed with Pi, e, and/or Phi. In other words, it is possibly all-pervasive in the human race. The very physical significance of zero, that is, nothingness in the background of nonnothingness, is in-built among all human beings and also perhaps among any living being. Clearly not much historical written/published record would be available to us—the modern men—as zero has continued to remain in the mental plane much more than in physical plane in the form of chronological exploration of this number/concept zero, central to all other numbers—negative and positive—existing from time immemorial to this date.

Furthermore, it is not difficult to imagine logically that many innovations on, about, and surrounding the zero in the world had taken place independently (without the knowledge of one another's contributions) and in parallel as the publication machinery and communication system were too poor in the olden days, and also natural calamities had taken their toll in terms of erasing considerably whatever that existed in some physical form or other.

Thus anybody who writes a history of zero starting from the remote past and ending today could at best write based on a meagre information available in printed literature and/or in symbols carved out on a stone slab and/or on imperishable/long-lasting objects. These objects were specifically constructed by human beings to denote intended information that withstood the onslaught of time and natural/artificial disasters such as earthquakes, hurricanes, forest fires, and wars. Consequently, such a history will remain very much incomplete and would fall short of our complete satisfaction.

Based on certain inscriptions and/or physical objects discovered centuries later a scientist/mathematical historian could draw inferences using his own (interpolation/extrapolation based) logic/knowledge about zero. Such inferences are often done and are sometimes (not always though) widely accepted and become useful in constructing the desired links of the past and the present. Since no writer of the history of mathematics has any other better alternative to write about the history of zero, his writing has a limited scope but would still be of interest to the reader, when it comes to the question of landmark innovations.

It is mostly immaterial if the concerned credit goes to an individual or a group and/or a country so far as the knowledge part of an innovation and its benefit to mankind are concerned. Under these circumstances, we venture to put forth a history of zero, that we believe would provide some significant insight about how the minds of our forefathers have worked to advance the frontier of knowledge that we are fortunate enough to be equipped with and benefited from it today.

3.1.2 Aryabhatta: use of decimals, zero, and place value system

The *origin of the modern decimal-based place value notation*, as mentioned earlier, is due to *Aryabhatta* in the third millennium BC. He provided elegant results for the summation of series of squares and cubes and made use of decimals, the zero (sunya), and the place value system. A decimal system was already in place in India during the *Harappan period*, which emerged before 2600 BC along the Indus River valley, as indicated by an analysis of Harappan weights and measures. In fact, weights corresponding to ratios of 0.05, 0.1, 0.2, 0.5, 1, 2, 5, 10, 20, 50, 100, 200, and 500 have been identified.

To find an approximate value of π, Aryabhatta gave the following prescription: Add 4 to 100, multiply by 8, and add to 62,000. This is "approximately" the circumference of a circle whose diameter is 20,000. This means that $\pi = 62,832/20,000 = 3.1416$. It is important to note that Aryabhatta used the word asanna (approaching), to mean that not only is this an approximation of π, but that the value is incommensurable or irrational (needing infinity of digits including zeros for its representation). This goes to show that zero was very much an integral part of a number—rational or irrational. The modern day zero is exactly the same (in meaning as well as in arithmetic) as the one used by Aryabhatta about 5000 years ago.

According to calculations of Aryabhatta (in *Surya Siddhanta*) the Hindu Kali Yuga (the fourth of four yugas or, ages in the scriptures of Hinduism) began at midnight (00:00) on February 18, 3102 BC and is currently continuing. Consequently, Aryabhatta dates the events of the (epic) Mahabharata to around 3137 BC.

3.1.3 The Maya numbers and Long Count

The *Maya* calendar dates the creation of the Earth to August 11 or August 13, 3114 BC establishing that date as day zero of the *Long Count* 13.0.0.0.0 (the Long Count days were tallied in a *modified base-20* (vigesimal) number scheme. Thus 0.0.0.1.3 is equal to 23, and 0.0.0.4.0 is equal to 80. The Long Count is not pure base 20, however, since the second digit from the right rolls over to zero when it reaches 18. Thus 0.0.1.0.0 does not represent 400 days, but rather only 360 days and 0.0.0.17.18 represents 358 days).

Although zero became an integral part of Maya numbers, with a different, empty tortoise-like "shell shape" employed for many depictions of the number of "zero," it did not influence Old World number systems *Quipu*, a knotted cord device, used in the *Inca* Empire and its predecessor societies in the Andean region to record accounting and other digital data. It is encoded in radix 10 positional number system. Zero is represented by the absence of a knot in the appropriate position of Quipu.

3.2 2000 BC–1000 AD: zero reached its full development along with representation and arithmetic operations

During this 3000-year period lots of activities in a number system were made in the quest of a number system that truly could be universally usable for the human progress in all sciences, engineering, accounts, finances, and economics/econometrics. The central issue was zero on which countless hours had been devoted by superminds of many subject areas including mathematics. Here we describe the events not chronologically but keeping in view the continuity of the development of the events. This is because several activities took place without much prior knowledge of all that had taken place elsewhere.

3.2.1 Representation of nothingness—an important need toward progress

How do we represent/express this state of universal or, equivalently, global nothingness? If we represent one (representing one object/item) by the symbol 1 in English, then should we express nothing by no symbol? Or should we introduce some distinct symbol different from all the symbols so far used to represent all nonnothing quantities as well as alphabetic and special characters? If nothing or, equivalently, a blank is used to represent "nothing," then confusion could be introduced with existing blanks all around.

A somewhat analogous situation was the representation of a point (i.e., "bindu" in Sanskrit and in many other Indian languages) or the definition of the point (a zero dimensional object). There is a need to represent a point on a paper. But for this, it would be difficult to convey its geometry to others. The point is defined as something that has no width, no length, and no height (in three-dimensional space), yet it exists. Such a contradictory definition is the only means to convey to the world what a point is.

In this context, it is worth noting that *Euclid* (325–265 BC) called a *geometrical object of zero dimension* "sign" (*semeion*), a practice which most successors followed, replacing the former word "point" (*stigme*). But the Roman scholars *Martianus Capella* (fifth century AD) and then *Boethius* (early sixth century AD) were to return to "point" (*punctum*) in their commentaries on Greek mathematics, and since then it has become standard. But "sign" is preferable; probably Euclid liked it since it emphasized the fact that signs are not subject to the operations of other geometrical magnitudes, which we now consider.

The Roman system of numerals used in ancient Rome (Ancient Rome was an Italic civilization, that is, an Indo-European ethno-linguistic group civilization that originated on the Italian Peninsula in the eighth century BC. Located along the coast of Mediterranean sea with Rome as its center, it expanded to become a largest empire over an area of 2.5 million square miles in the first–second century AD world with around 90 million inhabitants, which was approximately one fifth of the then world population) is still existing, a well-known use being the numbering of preliminary pages in a book.

The Romans, probably following their predecessors, the Etruscans, reacted against the Greek practice of using the letters of their alphabet in sequence for their integers. They, like the Greeks, had no zero. Their system was *not place-valued but was order-valued*, with the letters normally ordered in the forward direction of their written language. Contrary to later European practice, they rarely placed a lower-value numeral in front of a higher-value one to represent a subtraction, and so avoided zigzags such as "DCMCVILX."

Romans, as part of their administrative exactitude, developed a *refined* calendar (believed to have been a lunar calendar whose original version was attributed by the Roman writers to *Romulus*, the founder of Rome around 753 BC) in which *Calends* (Kalends) were the first days of each ancient Roman month. Unlike currently used dates, which are numbered sequentially from the beginning of the month, the Romans counted backwards from three fixed points: the *Nones*, the *Ides*, and the *Calends* of the following month.

Months were grouped in days such that the Calends (Kalends) were the 1st days of each ancient Roman month, the Ides were the 13th days of each short month or the 15th days of each long month. The Nones were 8 days before Ides and fell on the 5th or the 7th days of the month depending on the position of the Ides (Nones implies 9th since counting Ides as the 1st, 1 day before is the 2nd, and 8 days before is the 9th).

To find the day of the calends of the current month, one counts how many days remain in the month, and add 2 to that number. For example, April 25, is the 7th day of the calends of May, because there are 5 days left in April, to which 2 being added, the sum is 7. In other words, any day was named by counting backward from the next epochal day (the 11th day before the calends of the next month, for instance) in perplexing contrast to their *avoidance of subtraction in arithmetic*. Consistently with their *zero-free system*, the counting started from the day of counting—for them Tuesday was 4 days before Friday, not 3 days.

Also, how do we express local nothingness? Should we distinguish between the universal nothingness and local nothingness? Is the universal nothingness absolute? Is the local nothingness then nonabsolute or, equivalently, relative? How do we integrate these nothingnesses with things which already have some positive quantitative representation that can be easily understood? These have been pondered over by many—known and unknown scientists/mathematicians—over millennia. Different ideas have been put forth by mathematical/computational scientists in different parts of the world at different times or at the same time independently.

Also, several people—mathematicians/mathematical historians/nonmathematicians—applied their logic and ideas based on the then knowledge of zero which they had, interpolated (implying both interpolation and extrapolation), and brought to the world their own import of knowledge and ideas on zero.

It may be pointed out that the recording, publication, and communication machineries were significantly poor compared to what we have today. Consequently, some information that we see today and that have been reported in the literature need to be appropriately modified or pruned to arrive at more plausible/genuine information on zero. We address these issues along with the history/background and the final understanding of zero (in mathematics) resulting in the development of *calculus* (whose immense impact in all sciences, engineering, economics, business management/administration, and other areas of great practical importance is well-known and needs no special emphasis) over the last five millennia in the subsequent subsections.

3.2.2 Zero as a number used by Indians

There was a zero-like concept in ancient Mesopotamia, but for the Mesopotamians, zero was not a separate number. Instead of a separate number, they used a space for zero. The Mayans were another discoverer of zero in another form, but unfortunately, they had no communications with the rest of the world, so their zero could not spread worldwide.

It were the Indian mathematicians who first used zero as a number, and used a *circle* for it. Later, the Arabs adopted the Indian zero and used it in their mathematics. From Arabs, zero went to Europeans and then it spread worldwide.

3.2.2.1 Bhaskara II's Siddhanta Siromani: used zero of today

No doubt that the Indian mathematicians first used zero in its present form, in concept and as a separate number. *Bhaskaracharya* (*Bhaskara II*, working 486 AD, the son of Chudamani Maheshvar, was born in Bijapur district, Karnataka, India. He was the head of the astronomical observatory at Ujjain. Bhaskaracharya's celebrated work *Siddhanta Siromani* (Crown of Treatises) consists of four parts, namely, Leelavati,

Bijaganitam, Grahaganitam, and Goladhyaya. The first two deal exclusively with mathematics and the last two address astronomy), a great mathematician of fifth century AD was credited to have used the zero that we know of today.

3.2.2.2 *Sarvanandi's Lokavibhaga: reference to zero in Jain work*

A contemporary reference to zero is in a Jain work (a subset of Indian mathematics) *Lokavibhâga* ("Parts of the Universe") which possesses the very exact date of Monday, August 25th of the year 458 CE in the Julian calendar. It is the oldest Indian text known to be in existence which contains zero and the place-value system expressed in numerical symbols. This was written (in 458 CE) by a Jain ascetic *Acharya Sarvanandi*. Lokavibhâga is a book on Jain Cosmology. It was written in the rein of the *Pallav* King *Sinhvarman*, at the city of *Patalika* in *Vanrashtra* (probably near river Yamuna). The book was written in *Prakrit* language. Later the book was translated to *Sanskrit* by another Jain *Acharya Sinhasuri*. The Sanskrit translation is available.

It is notable that Bhaskaracharya himself brought the numbers to his Sanskrit works from the numbers which already existed in *Bramhi script* (Brāhmī is an oldest writing system used in the Indian subcontinent and in Central Asia during the final centuries BCE and the early centuries CE. Like its contemporary, *Kharosthi*, which was used in what is now Afghanistan and Western Pakistan, Brahmi (native to north and central India) was an abugida—a segmental writing system in which consonant-vowel sequences are written as a unit: each unit is based on a consonant letter, and vowel notation is secondary. The best-known Brahmi inscriptions are the rock-cut edicts of Ashoka in north-central India, dated to 250–232 BCE).

In ancient India, Brahmi script was exclusively used to write books and inscriptions in Prakrit languages. Ancient Prakrit languages and Brahmi script are closely related to Jain works. All the sacred text of Jains were written in Prakrit Languages and Brahmi script.

Bhaskara II, frequently uses zero and the place-value system, which are expressed in the form of numerical symbols. He also describes methods of calculation which are very similar to our own and are carried out using the nine numerals and zero. Moreover, he explains the fundamental rules of algebra where the zero is presented as a mathematical concept, and defines infinity as the inverse of zero.

3.2.2.3 *Sridhara's Patiganita, Ganitasara, and Ganitapanchavimashi: algorithms for arithmetic operations*

Sridhara (before 486 AD) was born perhaps in southern India or Bengal. Bhaskara II referred to him as a distinguished mathematician and quoted his work in several places. Sridhara gave mathematical formulas for many practical problems such as those involving rates of travel, filling a cistern, barter, simple interest, purchase and sale, mixtures, and ratios.

His *Patiganita* (Mathematics of Procedures) is considered as an advanced mathematical work. In it, all the procedures or, equivalently, algorithms for carrying out arithmetic operations are presented in verse form without giving any proof. Some parts of this book are devoted to arithmetic and geometric progressions with fractional numbers or terms, and formulas for the sums of certain finite series. His other remarkable works include *Ganitasara* (essence of Mathematics) and *Ganitapanchavimashi* (Mathematics in 25 verses). He provided the first correct formula (in India) for the volume of a truncated cone and that of a sphere.

3.2.2.4 Jinabhadra Gani's Brihatkshetrasamasa: expression for a 12-digit number

The works of Jinabhadra Gani, Indian arithmetician, who lived at the end of the sixth century notably include *Brihatkshetrasamasa*, where he gives an expression for the number 224, 400, 000, 000 in the simplified Sanskrit system using the place-value system.

3.2.2.5 Haridatta's Grahacharanibandhana: alphabatical positional number system

Haridatta (born 650 CE) was an astronomer-mathematician of Kerala, India, who is believed to be the promulgator of the *Parahita system* of astronomical computations. This system of computations is widely popular in Kerala and Tamil Nadu. According to legends, Haridatta promulgated the Parahita system on the occasion of the *Mamankam* held in the year 683 CE. Mamankam was a 12-yearly festival held in Thirunnavaya on the banks of the Bharathapuzha river. Notably, Haridatta's works include *Grahacharanibandhana*, in which he tells of the fruit of his invention: a system of numerical notation which uses the letters of the Indian alphabet. This is based on the place-value system and a zero (always expressed by one of two letters). This system is called *Katapayddi*: the first alphabetical positional number system.

3.2.2.6 Shankaracharya's Sharirakamimamsabhashya: reference of place-value system

Shankaracharya (Shankara, 700–750). (Shankaracharya is a commonly used title of heads of monasteries called *mathas* in the *Advaita Vedanta* tradition. The title derives from Adi Shankara, an eighth-century CE reformer of Hinduism.) A Hindu philosopher of the early eighth century, his works notably include *Sharirakamimamsabhashya* (Commentary on the Study of the Self), where there is a reference to the place-value system of the Indian numerals.

3.2.2.7 Lalla's Shishyadhividdhidatantra: usage of place-value system

Lalla (around 720–790 AD) was an Indian mathematician and astronomer who belonged to a family of astronomers. His most famous work *Shishyadhividdhidatantra* was in two volumes: *On the computation of the positions of the planets* and *On the sphere*. In this work there is abundant usage of the place-value system recorded by means of Sanskrit numerical symbols.

3.2.2.8 Shankaranarayana's Laghubhāskarīyavivaraṇa: Place-value system of Sanskrit numerical symbols

Shankaranarayana (840–900 CE) was an Indian astronomer and mathematician in the court of King *Sthanu Ravi Varman* (844–885 CE) of the Later Cheras in Kerala. He is believed to have established the first astronomical observatory in India at Kodungallur in Kerala. His most famous work was *Laghubhāskarīyavivaraṇa*, which was a commentary on the Laghubhāskarīya of Bhaskara I, which in turn is based on the work of Aryabhatta. The work which is written in 869 CE for the author writes in the text that it is written in the Shaka year 791, which translates to 869 CE, notably includes a text in which the place-value system of Sanskrit numerical symbols is used frequently. He also uses the *katapayadi* method invented by Haridatta as mentioned earlier.

3.2.2.9 Gotama Siddha's Kai yuan zhan jing: Symbol zero, place-value system, and Indian method of computation

Gautama (or *Gotama) Siddha (8th century)* (not to be confused with Gautama Buddha, i.e., Siddhartha Gautama) also known as *Qutan Xida*, Chinese Buddhist astronomer, astrologer, and compiler of Indian origin, edited a work on astronomy and astrology entitled *Kai yuan zhan jing* (718–729 CE), where he mentioned the symbol of zero, and described zero, the place-value system, and Indian methods of calculation. He is best remembered for leading the compilation of the *Treatise on Astrology of the Kaiyuan Era* during the *Tang Dynasty* (618–907 AD), an imperial dynasty of China.

He was born in Chang'an, and his family was originally from India, according to a tomb stele (or stela, i.e., a stone or wooden slab, generally taller than it is wide, erected as a monument, very often for funerary art) uncovered in 1977 in Xi'an. The Gautama family had probably settled in China over many generations, and might have been present in China prior even to the foundation of the Tang period. He was most notable for his translation of *Navagraha* calendar into Chinese. He also introduced Indian numerals with zero (○) in 718 AD in China as a replacement of counting rods.

A number of mathematical manuscripts of the Tang dynasty were discovered in western China. Some can be found among Dunhuang manuscripts preserved in the British Museum in London. Some of the calendar systems such as *Qiyaoli* (Seven Luminaries Calendar) and *Jiuzhili* (Nine Controllers Calendar) generated during the Tang period depicts the Indian influence. The nine Controllers in Hindu tradition consist of the seven Luminaries and the two imaginary invisible "planets" *Rahu* (Luohou) and *Ketu* (Jidu).

The Indian concept of large numbers and that of small numbers also came to China in the fourth century and in the mid-sixth century, respectively.

3.2.2.10 Govindasvamin's Bhaskariyabhasya: examples of place-value system with Sanskrit numerical symbols

Govindasvamin (800–860 CE). Indian astronomer. Notably, his works include *Bhaskariyabhasya*, in which there are many examples of the use of the place-value system using Sanskrit numerical symbols.

3.2.3 Egyptian number system

Ancient Egyptian numerals were *radix-10* (base-10) based. They used hieroglyphs for the digits and were not positional. By 1740 BCE the Egyptians had a symbol for zero in accounting texts. The symbol *nfr*, meaning beautiful, was also used to indicate the base level in drawings of tombs and pyramids and distances were measured relative to the base line as being above or below this line.

3.2.4 Babylonian number system

By the middle of the second millennium BC, the *Babylonian* mathematics had a sophisticated *sexagesimal* (base 60) positional number system. The lack of a positional value (or zero) was indicated by a *space* between sexagesimal numbers. By 300 BC, a punctuation symbol (two slanted wedges) was co-opted as a place holder in the same Babylonian number system. In a tablet unearthed at Kish dating from about 700 BC, the scribe *Bêl-bân-aplu* wrote his zeros with *3 hooks, rather than 2 slanted wedges*.

The Babylonian placeholder was not a true zero because it was not used alone. Nor was it used at the end of a number. Thus numbers like 2 and 120 (2×60), 3 and 180 (3×60), 4 and 240 (4×60), looked the same because the larger numbers lacked a final sexagesimal placeholder. Only context could differentiate them.

3.2.5 Greek number system

The *Greek* used a *base 10* number system drawing upon the letters of their alphabet (that included three letters that have since been discarded) during 500 BC. For clarity/nonambiguity, they sometimes used an over-bar or prime-stroke: thus α' for 1, …, θ' for 9, τ' for 10, …, κ' for 20, and so on. They do not seem to have used a letter/symbol exclusively for zero.

3.2.6 Bhaskara I's Aryabhatteeyabhashya: oldest Sanskrit prose work on mathematics

Bhāskara (before 123 BC) commonly called *Bhaskara I* to avoid confusion with the fifth century Indian mathematician Bhāskara II, working 486 AD) was born at Bori, in Parbhani district of Maharashtra state in India. He is the earliest known commentator of Aryabhatta's works. Bhaskara I was the first to write numbers in the Hindu–Arabic decimal system with a *circle for the zero*, and gave a unique and remarkable rational approximation of the sine function in his commentary on Aryabhatta's work.

This commentary, *Aryabhatteeyabhashya*, is the oldest known prose work in Sanskrit on mathematics and astronomy. He also wrote two astronomical works in the line of Aryabhatta's school, the *Mahabhaskariya* and the *Laghubhaskariya*.

We now come to considering the first appearance of zero as a number and its role in arithmetic operations. Let us first note that it is not in any sense a natural candidate for a number. Since early times *numbers are words* which refer to collections of objects. Certainly the idea of number became more and more abstract and this abstraction then made possible the consideration of zero and negative numbers which do not crop up as properties of collections of objects. Of course the question that arises when one attempts to consider zero and negatives as numbers is how they interact in regard to the arithmetic operations viz. addition, subtraction, multiplication, and division. In three important books the Indian mathematicians Brahmagupta (born 30 BC), Bhaskara (*Bhaskara II*, working 486 AD), and *Mahavira* (815–878 AD), dealt with this question.

3.2.7 Brahmagupta's Brahmasputa Siddhanta: understanding role of zero and computation

Brahmagupta's best known work, the *Brahmasputa Siddhanta* (Correctly Established Doctrine of Brahma), was written in Bhinmal, a town in the Jalore district of Rajasthan, India. Its 25 chapters contain several unprecedented mathematical results. It contains ideas including a good understanding of the mathematical role of zero, rules for manipulating both negative and positive numbers, a method for computing square roots, methods of solving linear and some quadratic equations, and rules for summing series, the Brahmagupta's identity, and the Brahmagupta's theorem.

He was the first to use zero as a number. He gave rules to compute with zero. Besides positive numbers, he used negative numbers and zero for computing. The modern rule that two negative numbers multiplied together equals a positive number first appears in *Brahmasputa Siddhanta*. It is composed in elliptic verse, as was common practice in Indian mathematics, and consequently has a poetic ring to it. As no mathematical proofs are given, it is not known how Brahmagupta's mathematics was derived. He attempted to give the rules for arithmetic involving zero and negative numbers in the early first century AD.

3.2.8 Brahmagupta's rules to compute with zero and later activities

The rules governing the use of zero that appeared for the first time in Brahmagupta's book Brahmasputha Siddhanta and considered not only zero, but negative numbers, and the algebraic rules for the arithmetic operations with such numbers also are truly outstanding during the very beginning of first century AD (over 2000 years ago). In some instances, his rules (stated below) differ from the modern standard:

1. The sum of zero and a negative number is negative.
2. The sum of zero and a positive number is positive.
3. The sum of zero and zero is zero.

4. The sum of a positive and a negative is their difference; or, if their absolute values are equal, zero.
5. A positive or negative number when divided by zero is a fraction with the zero as denominator.
6. Zero divided by a negative or positive number is either zero or is expressed as a fraction with zero as numerator and the finite quantity as denominator.
7. Zero divided by zero is zero.

In saying zero divided by zero is zero, Brahmagupta differs from the modern position. Mathematicians normally do not assign a value to this, whereas computers and calculators sometimes assign NaN, which means "not a number." Moreover, nonzero positive or negative numbers when divided by zero are either assigned no value, or a value of unsigned infinity, positive infinity, or negative infinity. It may be noted that *in natural mathematics* (mathematics that nature performs without any error) that follows all the laws of nature perfectly without any violation of any law, *division by zero can never occur since it implies a serious violation of a law of nature.*

He explained that given a number, if you subtract it from itself you obtain zero.

Brahmagupta gave the following rules for subtraction which is considered slightly harder.

A negative number subtracted from zero is positive, a positive number subtracted from zero is negative, zero subtracted from a negative number is negative, zero subtracted from a positive number is positive, zero subtracted from zero is zero.

So far so good as we know today regarding addition and subtraction. The foregoing rules stood the test of time and logic of all concerned and remain invariant (unchanged) so far and will remain so eternally.

Brahmagupta then says that any number when multiplied by zero is zero. This is also fine as we know today. However, it is believed that he struggles when it comes to division. When a division is viewed as the complementary operation of multiplication, zero in the operations poses a problem. The *physical significance* of division of a number by zero cannot be interpreted as anything in the physical world. The most important law not only in mathematics but also anywhere else (in the physical world/nature), that is, "Thou shalt not divide by zero" was introduced much later, but it is not known exactly when and (first) by whom. Probably several mathematicians have propounded this vital law independently at different times or at about the same time.

A positive or negative number when divided by zero is a fraction with the zero as denominator. Zero divided by a negative or positive number is either zero or is expressed as a fraction with zero as numerator and the finite quantity as denominator. Zero divided by zero is zero. The last statement, "Zero divided by zero is zero," as we know today is not valid if both the zeros are exact (or mathematical) zeros. As a matter of fact, mathematics considers that the division of zero by zero is undefined and in computing, it may sometimes be termed as NaN (not a number) as found in Matlab implementation. *However, it may not be viewed as too horrible or ridiculous a conclusion by Brahmagupta!*

Specifically there are computational mathematicians or even mathematicians who would not like to totally set aside the conclusion and would ponder over why Brahmagupta, an outstanding mathematician of the first century AD, passed such a statement. Was it so foolish of him? Was the statement so ridiculus/offensive? Consider the Newton's (*Sir Isaac Newton*, 1642–1727 AD) fixed-point second order iteration scheme (1) (developed during the seventeenth century and described later in this section) for computing a repeated root of the polynomial equation $f(x) = x^2 - 6x + 9 = 0$.

We recall that all man-made digital computers so far available and those that will be available in the future, possibly eternally, are finite precision (word-length) machines (irrespective of any technology—silicon or quantum computing, for example) and will never be infinite precision machines like the natural computer (a computing device (performing numerical, semi-numerical, and nonnumerical computations) which nature has been using since eternity and which is generally out of bounds for a human being) which works with actual real numbers (including those which have an infinity of digits in its representation) and knows no error.

It is well-known that the scheme will *oscillate* around the repeated zero 3 for any finite initial approximation and for sufficient number of iterations. In fact, the fixed-point iteration scheme is oscillatory (and nondivergent) for a multiple (repeated) root. It can be checked that increasingly the larger the precision of the man-made digital computer is, increasingly the smaller will be the relative error in the computed root 3, that is, the amplitude of the oscillation will be increasingly shorter.

Or, in other words, *the term $-f(x_i)/f'(x_i)$ in the scheme will be treated as a numerical zero for sufficiently large iteration number i*. This is because, in the term $-f(x_i)/f'(x_i)$ of the scheme, $f(x_i)$ tends to zero faster than its derivative $f'(x_i)$ for sufficiently large number of iterations (for a repeated root). *At the exact root viz. 3, both $f(x_i)$ and $f'(x_i)$ are exactly zero* while in numerical computation (which is always with finite precision) involving *numerical zeros* by the scheme, exact 0/0 may not arise at all in a digital computer with the widely used twenty-first century *finite precision* floating-point arithmetic.

It may be remarked that in physics/any natural science, in the realm of any continuous quantity (or even in the realm of a discrete quantity in which the number of distinct constituents is very/too large such as the number of red blood cells in one cubic milliliter), exact quantity is not known; a narrow bound at best depending on the measuring device is known in which the exact quantity lies. This is just to stress the fact that *in sciences, the term "zero" almost always refers to a "numerical zero" and not the exact (absolute) zero.*

Really Brahmagupta is saying very little when he suggests that *n* divided by zero is *n*/0. He may have struggled here. However it is a brilliant attempt from the first person that we know who tried to extend arithmetic to negative numbers and zero.

3.2.9 Bhaskara II's Siddhanta Siromani: writing on division by zero and Rolle's theorem

Bhaskara II (Bhāskarāchārya implying "Bhāskara the teacher," working 486 AD) wrote about 500 year after Brahmagupta regarding division by zero. Despite the passage of time he seemed to be still struggling to explain division by zero. He writes

> *A quantity divided by zero becomes a fraction the denominator of which is zero. This fraction is termed an infinite quantity. In this quantity consisting of that which has zero for its divisor, there is no alteration, though many may be inserted*

or extracted; as no change takes place in the infinite and immutable God when worlds are created or destroyed, though numerous orders of beings are absorbed or put forth.

So Bhaskara II tried to solve the problem by writing $n/0 = \infty$. At first sight we might be tempted to believe that Bhaskara has it correct, but of course he does not. If this were true then 0 times ∞ must be equal to every number n, so all numbers are equal. It appears that the fact "Thou shalt not divide by zero" was yet to occur or be firmly stated by the then Indian mathematicians although they were hovering close to it!

Bhaskara did correctly state other properties of zero, however, such as $0^2 = 0$, and $\sqrt{0} = 0$. We will see later that he depicted remarkable insight regarding the knowledge of zero. He gave a statement of *Rolle's theorem*, concluded that the derivative vanishes at a maximum, and introduced the concept of the instantaneous motion of a planet in his collection *Siddhanta Siromani* (Crown of treatises). *Probably the exact zero and a numerical zero could have been the cause for confusion as both have distinct roles in computation (specifically with reference to our modern digital computer that was nonexistent during the fifth century AD (Bhaskara II's time).*

3.2.10 Jain text Lokavibhaga: decimal place-value system, infinitive universe, big calculation

The Jain text *Lokavibhâga* (458 AD), mentioned earlier in this section, based on a decimal place-value system, employed *shunya* ("void" or "empty") in computation. In simple words, Jain mathematics could be described as the Mathematics developed and used by the Jain ascetics of India. Jain mathematicians contributed a lot for the development of mathematics in ancient times.

As Jain philosophy says that the *universe (in size and shape) is infinitive*—no beginning and no end, they did lot of research in the concepts of Space, Time and Matter. One may contrast this with the *Big Bang theory*. Jain mathematicians divided large numbers in three categories—*Enumerable* (Countable), *Innumerable* (Not countable), and *Infinitive* (Endless). According to them, all infinitives are not the same and consequently they divided infinitive into five categories.

It was a necessity of the Jain ascetics to make big calculations. All of this developed Jain mathematics. From the Jain cosmological book, we find the 178-digit number (computed here by us by Matlab command >>*vpa(2^588, 178)*, where vpa denotes *variable precision arithmetic*)

$$2^{588} = 1013\ 065324\ 433836\ 171511\ 818326\ 096474\ 890383\ 898005\ 918563\ 696288\ 002277\ 756507\ 034036\ 354527\ 929615\ 978746\ 851512\ 277392\ 062160\ 962106\ 733983\ 191180\ 520452\ 956027\ 069051\ 297354\ 415786\ 421338\ 721071\ 661056$$

which is believed to be the age of the universe in years. Besides these, Jain mathematicians worked on indices, set theory, permutation, combinations, and many other concepts we use today, which involve the use of the decimal number system that includes 0 as a number—the most important of all the digits in any number system.

3.2.11 Yajur Veda Samhita: numeral denominations

Nine numbers (1, 2, 3, 4, 5, 6, 7, 8, and 9) and a zero (0) can be combined to form infinite mathematical expressions and measurements. This knowledge is said to be the unique contribution of ancient Indian genius to the world's progress. During the Vedic era this decimal system was very much in vogue in India. The *2nd mantra of the 17th chapter of Yajur Veda Samhita* describes the numeral denominations:

Eka (1), *daśa* (10), *sata* (100), *sahasra* (1000), *ayuta* (10,000), *laksha* (100,000), *niyuta* (1,000,000), *koti* (10,000,000), …, *parardha*, ….

3.2.12 Buddhistic, Jain, and Greek texts: description of their largest numbers

A Buddhistic text *Lalitha Vistara* (first century BC) describes up to 10^{53} and called that numerical value as *Talakshna*. Another Jain text *Anuyogadwara* describes numbers up to 10^{140}. The ancient Greeks gave the biggest numerical value called *myriad* which is technically the number ten thousand (10^4); in that sense, the term is used almost exclusively in translations from Greek, Latin, or Chinese, or when talking about ancient Greek numbers. More generally, a myriad may be an indefinitely large number of things.

3.2.13 Stone/copper plate inscription in Gwalior: circle for digit zero

The first known use of special glyphs for the decimal digits that includes the indubitable appearance of a symbol for the digit zero, a *small circle*, appears on a stone inscription found at the Chaturbhuja Temple at *Gwalior* in India, dated 876 AD.

We have an inscription on a stone tablet that has a date which translates to 876 (AD). The inscription was found in Gwalior, a city in the state of Madhya Pradesh, India, located 122 kilometers (76 miles) south of Agra, 423 kilometers (263 miles) north of Bhopal, and 400 kilometers (249 miles) south of Delhi, where they planted a garden 187 hastas by 270 hastas. One hasta (1.5 feet) is a traditional Indian unit of length, measured from the elbow to the tip of the middle finger. This garden would produce enough flowers to allow 50 garlands per day to be delivered to the local temple. Both of the numbers 270 and 50 are denoted almost as they appear today although the 0 is smaller and slightly raised. There are many documents on copper plates, with the same small *o* in them, dated back as far as the *sixth century AD*, but their authenticity may be controversial.

3.2.14 al-Khwarizmi: Hindu–Arabic numerals and treatises on astronomy and algebra

To appreciate the contribution of *al-Khwarizmi* (Abu Jafar Mohammed Ibn Musa al-Khwarizmi (around 780–850), "Mohammed the father of Jafar and the son of Musa"), possibly the most celebrated mathematician and astronomer of ninth century

AD in the Arabic and Persian world, it would be interesting and helpful to know the environment of his time including the royal patronage/encouragement and appreciation received by him. He was born to a *Persian* family perhaps in Khwarezm (Khiva), Uzbekistan; however, the epithet *al-Qutrubbulli* indicates he might be from Qutrubbull, a small town near Baghdad.

Gerald James Toomer (born 1934 AD) indicated that he was an adherent of the old Zoroastrian religion, but the pious preface to al-Khwarizmi's Algebra shows that he was an orthodox Muslim (true believer). In 786 AD, Harun al-Rashid (766–809 AD) became the fifth Caliph of the Abbasid dynasty, and ruled from his court in the capital city of Baghdad over the Islamic empire, which stretched from the Mediterranean to India. He brought culture to his court and tried to establish the intellectual disciplines, which at that time were not flourishing in the Arabic world.

He had two sons, the younger was al-Amin (787–813 AD), and the elder was al-Mamun (786–833 AD). Harun died in 809 AD and there was an armed conflict between his sons. al-Mamun won the armed struggle, and al-Amin was defeated and killed in 813 AD. al-Mamun became the new Caliph and ruled the empire from Baghdad, continuing the *patronage of learning* started by his father, who was said to have been inspired by a dream in which Aristotle appeared to him.

al-Mamun founded an academy called the *House of Wisdom* (*Bait al-hikma*). He also built up a *library of manuscripts*, the first major library since the library at *Alexandria*, that collected important works from Byzantium. In addition to the House of Wisdom, al-Mamun set up observatories in which Muslim astronomers could build on the knowledge acquired by earlier peoples. In *Baghdad*, scholars encountered and built upon the ideas of ancient Greek and Indian mathematicians.

al-Khwarizmi, al-Kindi, Banu Musa (around 800–860 AD), Ibrahim al-Fazari (died 777), his son Mohammad al-Fazari (died 806 AD), and Yaqub ibn Tariq (died about 796 AD) were scholars at the House of Wisdom in Baghdad. Their tasks there involved the translation of Greek and Sanskrit scientific manuscripts. They also studied and wrote on algebra, geometry, and astronomy. There, al-Khwarizmi encountered the Hindu place-value system based on the numerals 0, 1, 2, 3, 4, 5, 6, 7, 8, 9, including the first use of zero as a place holder in positional base notation, and he wrote a treatise around 820 AD on what we call Hindu–Arabic numerals.

The Arabic text is lost, but a Latin translation, *Algoritmi de numero Indorum*, (that is, al-Khwarizmi on the Hindu Art of Reckoning), a name given to the work by Baldassarre Boncompagni (1821–1894 AD) in 1857 AD, changed much from al-Khwarizmi's original text, of which even the title is unknown. This book synthesized Greek and Hindu knowledge and also contained his own fundamental contribution to mathematics and science including an explanation of the use of zero. It was only centuries later, in the twelfth century, that the Arabic numeral system was introduced to the Western world through Latin translations of his treatise *Arithmetic*.

However, in his Sefer ha mispar (Number Book), Rabbi Ben Ezra (1092–1167) used the term sifra. In various spelling, the Arabic term sifra (cifra, cyfra, cyphra, zyphra, tzphra, …) continued to be used to mean "zero" by some mathematicians for many centuries: we find it in the Psephophoria kata Indos (Methods of Reckoning of the Indians) by the Byzantine monk Maximus Planudes (1260–1310) in the

Institutiones mathematicae of Laurembergus, published in 1636, and even as late as 1801 in Karl Friedrich Gauss's Disquisitiones arithmeticae (Gauss must have been one of the very last scholars to write in Latin).

The French Minorite friar Alexander de Villa Dei, who taught in Paris around 1240 AD, mentions an Indian King named Algor as the inventor of the new "art," which itself is called the algorismus. Thus the word "algorithm" was tortuously derived from al-Khwarizmi (Alchwarizmi, al-Karismi, Algoritmi, Algorismi, Algorithm), and has remained in use to this day in the sense of an arithmetic operation. This Latin translation was crucial in the introduction of Hindu–Arabic numerals to medieval Europe.

al-Khwarizmi dedicated two of his texts to the Caliph al-Mamun. These were his treatises on algebra and astronomy. The algebra treatise, *Hisab al-jabr w'al-muqabala*, was the most famous and important of all of al-Khwarizmi's works (the Arabic word al-jabr means setting of a broken bone/restoration/completion). al-Khwarizmi's original book on algebra is lost. It is the Latin translation of the title of this text, *Liber algebrae et almucabala*, in 1140 AD by the Englishman Robert of Chester (about 1150 AD), and by the Spanish Jew John of Seville (about 1125 AD) that gives us the word "algebra."

This book was intended to be highly practical, and algebra was introduced to solve problems that were part of everyday life in the Islamic empire at that time, such as those that men constantly encounter in cases of inheritance, legacies, partition, lawsuits, and trade, and in all their dealings with one another, or where the measuring of lands, the digging of canals, geometrical computations, and other objects of various sorts and kinds are concerned.

Early in the book al-Khwarizmi describes the natural numbers, and then introduces the main topic of the first section of his book, namely, the solution of equations with a single unknown. His equations are linear or quadratic and are composed of units, roots, and squares. al-Khwarizmi's mathematics was done *entirely in words, no symbols were used*. A unique Arabic copy of this book is kept at Oxford and was translated in 1831 AD by F. Rosen. A Latin translation is kept in Cambridge. Western Europeans first learned about algebra from his works.

In 773, a man well versed in astronomy by the name of *Kanaka* from India brought with him the writings on astronomy, the *Siddhanta* of Brahmagupta. Around 820 AD, during the rule of Caliph al-Mamun (786–833 AD) who continued the patronage of learning started by his predecessors (specially at the behest of Caliph al-Mansur (Abu Ja'far Abdallah ibn Muhammad Al-Mansur, 714–775 AD), the second Abbasid Caliph), translations were made of many ancient treatises including Greek, Latin, Indian, and others. al-Khwarizmi was asked to translate this work from Sanskrit into Arabic, which became known as the *Sindhind*. It was promptly disseminated, and induced Arab scholars to pursue their own investigations into astronomy. The original Arabic version of Sindhind is lost, however, a Latin translation by Adelard of Bath (around 1080–1152 AD) in 1126 AD, which is based on the revision by *al-Majriti* (about 950 AD), has survived.

The four surviving copies of the Latin translation are kept at the Bibliothéque Publique (Chartres), the Bibliothéque Mazarine (Paris), the Bibliotheca Nacional

(Madrid), and the Bodleian Library (Oxford). This work consists of approximately 37 chapters and 116 tables on calendars; calculating true positions of the Sun, Moon, and planets; tables of sines and tangents; spherical astronomy; astrological tables; parallax and eclipse calculations; and visibility of the Moon.

al-Khwarizmi also wrote a longer version of Sindhind, which is also lost, but a Latin version has survived. Some historians assume that al-Khwarizmi was influenced by Ptolemy's work in casting tables in Sindhind; however, he did revise and update Ptolemy's Geography, and authored several works on astronomy and astrology. The book *Kitab surat al-ard* (The Image of the Earth), which is based on Ptolemy's Geography, lists coordinates of latitudes and longitudes for 2402 cities in order of "weather zones," mountains, seas, islands, geographical regions, and rivers, but neither the Arabic copy nor the Latin translation include the map of the world. A number of other works were also written by al-Khwarizmi on the astrolabe, sundial, and the Jewish calendar.

Several Arabic manuscripts in Berlin, Istanbul, Tashkent, Cairo, and Paris contain further material that perhaps comes from al-Khwarizmi. He also wrote a political history containing horoscopes of prominent persons. According to one story, he was called to the bedside of a seriously ill caliph and was asked to cast his horoscope. He assured the patient that he was destined to live another 50 years, but caliph died within 10 days. A postage stamp was issued by the USSR in 1983 to commemorate the 1200th anniversary of Muhammad al-Khwarizmi. The foregoing vast contributions of al-Khwarizmi demonstrate the use of full-fledged decimal number system with all the five attributes that today's (Indian) zero has.

3.2.15 Mahavira's text Ganita Sara Samgraha: update of Brahmagupta's book

In 830 AD, around 800 years after Brahmagupta wrote his masterpiece, Mahavira (or Mahaviracharya meaning *Mahavira* the Teacher born in Mysore or a nearby place in Southern India during 817 AD and passed away in 875 AD) wrote *Ganita Sara Samgraha*, dated 850 AD, which was designed as an update of Brahmagupta's book. He correctly states ... a number multiplied by zero is zero, and a number remains the same when zero is subtracted from it. Mahavira was of the Jaina (Jain) religion and was familiar with Jaina mathematics. He worked in Mysore where he was a member of a school of mathematics. However A. Jain mentions six other works which he credits to Mahavira and he emphasizes the need for further research into identifying the complete list of his works.

3.2.16 Chinese counting rods: zero as a vacant position not a number

The *Chinese* text *Sunzi Suanjing* of unknown date but estimated to be dated from the first to fifth centuries, and Japanese records dated from the eighteenth century AD, describe how Chinese *counting rods* were used for calculations. According to *A History of Mathematics*, the rods "gave the decimal representation of a number, with an empty space denoting zero." The counting rod system is considered a positional notation system.

Zero was not treated as a number at that time, but as a "vacant position," unlike the Indian mathematicians who developed the zero as a number. *Ch'in Chu-shao's* 1247 AD *Mathematical Treatise in Nine Sections* is the oldest surviving Chinese mathematical text using a round symbol for zero. Chinese authors had been familiar with the idea of negative numbers by the Han Dynasty (second century CE), as seen in the *The Nine Chapters on the Mathematical Art* much earlier than the fifteenth century when they became well established in Europe.

3.2.17 Ancient Greeks: how can nothing be something and Zeno's paradoxes

Records show that the *ancient Greeks* seemed unsure about the status of zero as a number. They asked themselves, "*How can nothing be something*?" leading to philosophical and, by the Medieval period, religious arguments about the nature and existence of zero and the vacuum. The *paradoxes of Zeno* of Elea (a set of philosophical problems generally thought to have been devised by Greek philosopher Zeno of Elea, 490–430 BC. One of his most famous paradoxes is that of motion, that is, "*In a race, the quickest runner can never overtake the slowest, since the pursuer must first reach the point whence the pursued started, so that the slower must always hold a lead*") depend in large part on the uncertain interpretation of zero.

3.2.18 Ptolemy: Hellenistic zero, sexagesimal numeral system, and Almagest

By 130 AD, *Claudius Ptolemy*, influenced by Hipparchus and the Babylonians, was using a symbol for zero (a small circle with a long overbar) within a *sexagesimal numeral system* otherwise using alphabetic Greek numerals. Because it was used alone, not just as a placeholder, this *Hellenistic zero* was perhaps the first documented use of a *number* zero in the Old World. However, the positions were usually limited to the fractional part of a number (called minutes, seconds, thirds, fourths, etc.)—they were not used for the integral part of a number. In later Byzantine manuscripts of Ptolemy's *Syntaxis Mathematica* (also known as the *Almagest*), the Hellenistic zero had morphed into the Greek letter omicron (otherwise meaning 70).

3.2.19 Bede: Roman numerals and zero

Another zero was used in tables alongside *Roman* numerals by 525 AD (first known use by Dionysius Exiguus), but as a word, *nulla* meaning "nothing," not as a symbol. When division produced zero as a remainder, *nihil*, also meaning "nothing," was used. These medieval zeros were used by all future medieval computists (calculators of Easter). The initial "N" was used as a zero symbol in a table of Roman numerals by the English monk Bede (672/673–735 AD), also referred to as *Saint Bede* or the *Venerable Bede* or his colleague, around 725 AD.

3.2.20 Zero (0): Placeholder, driver of calculus, and most pervasive global symbol

When anyone thinks of one hundred, two hundred, or seven thousand, the image in his or her mind is of a digit followed by a few zeros. The zero functions as a placeholder; that is, seven followed by three zeroes denotes that there are seven thousands, rather than only seven hundreds if seven is followed by only two zeros. If we were missing one zero, that would drastically change the amount. Just imagine having one zero erased (or added) to your salary! Yet, the number system we use today—Arabic, though it in fact came originally from India—is relatively new. For centuries people marked quantities with a variety of symbols and figures, although it was awkward to perform the simplest arithmetic calculations with these number systems.

From placeholder to the driver of calculus, zero has crossed the greatest minds and most diverse borders since it was possibly born over seven millennia ago. Today, *zero (0) is perhaps the most pervasive global symbol known.*

Understanding and working with zero is the basis of our world today; without zero we would lack calculus, financial accounting, the ability to make arithmetic computations quickly, and, especially in today's connected world, computers. The story of zero is the story of an idea that has aroused the imagination of super-minds across the globe.

3.2.21 Brahmagupta: arithmetic operations using zero and al-Khwarizmi's algebra and algorithms

Brahmagupta was the first to formalize arithmetic operations (discussed earlier in this section) using zero during *the late first century BC/early first century AD*. He wrote standard rules for reaching zero through addition and subtraction as well as the results of operations with zero. The only error in his rules was division by zero (rather mathematical zero as opposed to numerical/relative zero).

But it would still be a few centuries before zero reached Europe. First, the great Arabian voyagers would bring the texts of Brahmagupta and his colleagues back from India along with spices and other exotic items. Zero reached Baghdad by 773 AD and would be developed in the Middle East by Arabian mathematicians who would base their numbers on the Indian system. In the ninth century, *Mohammed ibn-Musa al-Khwarizmi* (around 780–850 AD) was the first to work on equations that equalled zero, or algebra as it has come to be known. He also developed quick methods for multiplying and dividing numbers known as algorithms (a corruption of his name). al-Khwarizmi called zero "*sifr*," from which our cipher is derived.

When the Arabs adopted Indian numerals and the zero, they called the latter sifr, meaning "empty," a plain translation of the Sanskrit shunya. Sifr is found in all Arabic manuscripts dealing with arithmetic and mathematics, and it refers unambiguously to the null figure in place-value numbering. Etymologically, sifr means "empty" and also "emptiness" (the later can also be expressed by khala, or faragh). The stem SFR can also be found in words meaning "to empty" (asfara),"to be empty" (safir), and "have–nothing" (safr al yadyn, literally "empty hands," that is to say, "he who has nothing in his hands."

By 879 AD, zero was written almost as we now know it, an oval—but in this case smaller than the other numbers. Thanks to the conquest of Spain by the Moors, zero finally reached Europe. By the middle of the twelfth century AD, translations of al-Khwarizmi's work had weaved their way to England.

3.3 1000–1900 AD: introduction of Hindu–Arabic numeral system in Europe

The *Hindu–Arabic numeral system* (radix 10) reached Europe in the eleventh century AD, via the Iberian Peninsula (commonly called Iberia, the third largest European peninsula (after the Scandinavian and Balkan peninsulas) which is located in the extreme southwest of the European continent. The area is approximately 225,000 sq miles. There are three countries in it (in addition to subterritories of France and Britain): Spain, Portugal, Andorra (a land-locked microstate—sixth smallest nation in Europe with an area of 181 sq miles and a population of less than 100,000 as in 2013 AD—in south-western Europe, located in the eastern Pyrenees mountains and bordered by Spain and France), as well as a part of France and the British Overseas Territory of Gibraltar) through Moors—the Spanish Muslims—together with knowledge of astronomy and instruments like the astrolabe, first imported by Gerbert of Aurillac. For this reason, the numerals came to be known in Europe as "*Arabic numerals*."

3.3.1 Fibonacci's Liber Abaci: introduction of Hindu numerals to Europe

The famed Italian mathematician, *Fibonacci*, built on Al-Khwarizmi's work with algorithms in his book *Liber Abaci*, or "Abacus book," in 1202 AD. Until that time, the abacus had been the most prevalent tool to perform arithmetic operations. Fibonacci's developments quickly gained notice by Italian merchants and German bankers, especially the use of zero. Accountants knew their books were balanced when the positive and negative amounts of their assets and liabilities equalled zero. But governments were still suspicious of Arabic numerals because of the ease in which it was possible to change one symbol into another. Though outlawed, merchants continued to use zero in encrypted messages, thus the derivation of the word cipher, meaning code, from the Arabic sifr.

Fibonacci was instrumental in bringing the system into European mathematics in 1202 AD, stating:

> *After my father's appointment by his homeland as state official in the customs house of Bugia for the Pisan merchants who thronged to it, he took charge; and in view of its future usefulness and convenience, had me in my boyhood come to him and there wanted me to devote myself to and be instructed in the study of calculation for some days. There, following my introduction, as a consequence of marvellous instruction in the art, to the nine digits of the Hindus, the knowledge*

of the art very much appealed to me before all others, and for it I realized that all its aspects were studied in Egypt, Syria, Greece, Sicily, and Provence, with their varying methods; and at these places thereafter, while on business. I pursued my study in depth and learned the give-and-take of disputation. But all this even, and the algorism, as well as the art of Pythagoras, I considered as almost a mistake in respect to the method of the Hindus (Modus Indorum). Therefore, embracing more stringently that method of the Hindus, and taking stricter pains in its study, while adding certain things from my own understanding and inserting also certain things from the niceties of Euclid's geometric art. I have striven to compose this book in its entirety as understandably as I could, dividing it into fifteen chapters. Almost everything which I have introduced I have displayed with exact proof, in order that those further seeking this knowledge, with its pre-eminent method, might be instructed, and further, in order that the Latin people might not be discovered to be without it, as they have been up to now. If I have perchance omitted anything more or less proper or necessary, I beg indulgence, since there is no one who is blameless and utterly provident in all things. The nine Indian figures are: 9 8 7 6 5 4 3 2 1. With these nine figures, and with the sign 0 … any number may be written.

Here Fibonacci uses the phrase "sign 0," indicating it is like a sign to carry out operations like addition or multiplication. From the thirteenth century AD, manuals on calculation (such as adding, multiplying, and extracting roots) became common in Europe where they were called *algorismus* after the Persian mathematician al-Khwārizmī.

3.3.2 Sacrobosco: defects of Julian calendar and recommendation on Gregorian calendar

The most popular introductory text to the Hindu–Arabic numeral system was written by *Johannes de Sacrobosco* (1195–1256 AD. He was a scholar, monk, an astronomer, and a teacher at the University of Paris. He wrote a short most widely read (during the late middle ages, i.e., thirteenth–fifteenth centuries) introduction to the Hindu–Arabic numeral system. He correctly described the defects of the then used Julian calendar and with a good degree of precision he recommended what was essentially the Gregorian calendar) in about 1235 AD and was one of the earliest scientific books to be *printed* in 1488 AD. Until the late fifteenth century AD, Hindu–Arabic numerals seem to have predominated among mathematicians, while merchants preferred to use the Roman numerals. In the sixteenth century AD, they became commonly used in Europe.

3.3.3 Shen Gua's Mengqi bitan: concept of infinitesimal and exhaustion Shen Gua (1031–1095 AD)

In the later half of the eleventh century, we come across the name of Shen Gua whose work *Mengqi bitan* (Dream Pool Essays) has something of an algebraic and geometric character. Shen Gua describes in it the art of piling up very small things, "*zowei zhi shu*." He used the term *qiji* (interstice-volumes) for volumes in discussing the outer indentations (*keque*) of and the empty interstices (*xuqi*) among a number of balls piled up together, or piled alterstep bricks (*cengtan*), or piled wine-kegs (*ying*). For areas, he uses the term *gehui zhi shu*, the art of cutting and making to meet. He seemed to know that the smaller the units become, the more fully it will be possible to exhaust the volume or area. He also used the term *zaige*, to cut repeatedly. Thus he came close to the concept of infinitesimal and exhaustion, this was never fully explored into the science of the calculus though.

3.3.4 Siefe and Kaplan on history of zero

The number zero as we know it arrived in the West by around 1200 AD, most famously delivered by Fibonacci who brought it, along with the rest of the Arabic numerals, back from his travels to north Africa. But the history of zero, both as a concept and a number, stretches far deeper into history—so deep, in fact, that its provenance is difficult to nail down.

Charles Seife is an American author, journalist, and professor. His first published book was *Zero: The Biography of a Dangerous Idea*. This book was his most famous work. Charles says "There are at least two discoveries, or inventions, of zero," "The one that we got the zero from came from the Fertile Crescent." The *Fertile Crescent* is a crescent-shaped region containing the comparatively moist and fertile land of otherwise arid and semi-arid Western Asia, the Nile Valley, and the Nile Delta of north-east Africa. The term was first used by archaeologist James Henry Breasted of University of Chicago. Having originated in the study of ancient history, the concept soon developed and today retains meanings in international geopolitics and diplomatic relations. It first came to be between 400 and 300 BC in Babylon, Charles says, "Before developing in India, wending its way through northern Africa and, in Fibonacci's hands, crossing into Europe via Italy."

Initially, zero functioned as a mere placeholder—a way to tell the difference of 1 from 10 and that of 10 from 100, to give an example using Arabic numerals. "That's not a full zero," Charles says. "A full zero is a number on its own; it's the average of -1 and 1." *It may be remarked that a blank (if assumed as a symbol) cannot take the place of zero because of inherent ambiguity as well as nonuniformity.*

Zero began to *take shape as a number, rather than a punctuation mark between numbers, in India in the fifth century AD*, says *Robert Kaplan*, "It isn't until then, and not even fully then, that zero gets full citizenship in the republic of numbers," Robert says. Some cultures were slow to accept the idea of zero, which for many carried darkly magical connotations.

The second appearance of zero occurred independently in the New World, in Mayan culture, likely in the first few centuries AD. Robert says, "That, I suppose, is the most striking example of the zero being devised wholly from scratch."

Robert pinpoints an even earlier emergence of a placeholder zero, a pair of angled wedges used by the Sumerians to denote an empty number column some 4000–5000 years ago.

But Charles is not certain that even a placeholder zero was in use so early in history. "I'm not entirely convinced," he says, "but it just shows it's not a clear-cut answer." He notes that the history of zero is too nebulous to clearly identify a lone progenitor. "In all the references I've read, there's always kind of an assumption that zero is already there," Charles says. "They're delving into it a little bit and maybe explaining the properties of this number, but they never claim to say, 'This is a concept that I'm bringing forth.'"

Robert's exploration of zero's genesis turned up a similarly blurred web of discovery and improvement. "I think there's no question that one can't claim it had a single origin," Robert says. "*Wherever you're going to get placeholder notation, it's inevitable that you're going to need some way to denote absence of a number.*"

In the well-known decimal number system consisting of 10 symbols, that is, 0, 1, 2, 3, 4, 5, 6, 7, 8, and 9 in English, the symbol 0 representing "nothing" or "empty" has been the most important innovation in any number system that we know and think of today. The name "zero" for the symbol 0 appears to have been derived ultimately from the Arabic *sifr* which also gives us the word "cipher."

One of the frequently asked queries is: Who discovered zero or the concept of zero? It is difficult to answer this question that logically satisfies the scientific community. It is not so important to know which individual genius had invented it. If somebody had come up with the concept of zero which everybody then perceived as a brilliant innovation that merits an entry into mathematics from that time on, the query would have a satisfactory answer.

The history, however, points to a different path leading to the concept of zero. Zero made shadowy appearances only to almost disappear again as if mathematicians were looking for it and yet failed to recognize its fundamental importance even when they saw it.

The first thing to say about zero is that there are at least two important uses of zero which are both extremely important but are somewhat different. One use is as an *empty place indicator* in the conventional place-value number system. Hence in a number like 4308 2106, the zero is used so that the positions of the digit 4 2 and the digit 3 are correct. Evidently, 438 216 denotes a number something quite totally different from the intended ones.

The other second use of zero is as a *mere number* itself in the form we use it as 0. There are also various different aspects of zero within these two uses, that is, the concept, the notation, and the name. Neither of the foregoing uses has a history that can be easily described and acceptable to all. It just did not happen that someone innovated the ideas, and then everyone started using them. Also it will not be wrong to say that the number zero is more than a mere intuitive concept. A mathematical problem (model) originated from a physical problem as a "real" problem rather than an abstract problem having no connection with the physical world.

Numbers in early historical times were thought of much more concretely than the abstract concepts which are our numbers today. It is not difficult to imagine that there was a *giant mental leap from 9 cows/horses to 9 "things" and then to the abstract idea of "nine."* If, in the historic/prehistoric era, one ancient people solved a problem about how many cows/horses a farmer needed then the problem was not going to have 0 or −15 as an answer.

The Muslims translated the Indian sunya as sifr. When Fibonacci wrote his Liber Abaci in 1202, he spoke of the symbol as Zephirum. A century later Maximus

Planudes (1340) called it Tzipha, and this form was still used as late as the sixteenth century. In Italian it was called Zenero, Cenero, and Zephiro. Since the fourteenth century, zero has been used as shown in the records of 1491 by Calnadri and of 1494 by Luca Pacisli. The word nulla appears in Italian translations of Muslim writings of the twelfth century, and also in the French (Nicolas Chuquet (1445 or 1455–1488 or 1500), a French mathematician invented his own notation for algebraic concepts and exponentiation. He may have been the first mathematician to recognize zero and negative numbers as exponents) and German writings of the fifteenth century.

Cipher was still used for zero by Adrian Metiers (1611), Herigone (1634), Cavalieri (1643), and Euler (1783), even though the more modern German word Ziffer had been introduced. The zero symbol also called cipher and naught, and modern usage permits it to be called 0 (letter "O")—an interesting return to the Greek name omicron. The Hindu symbol for the zero was circle with a dot at its center. However, in the Muslim Empire, the Muslim East (Asia, Baghdad) and the Muslim West (North Africa, Spain), a number of different representations were used. In the Muslim East, the dot "•" was used, where as the Muslim West adopted the circle ∘.

A convenient system of writing the numbers is a great asset in the development of mathematics. The use of zero and positional notation (i.e., writing all the numbers with the help of nine numerals and zero in units, tens, hundreds, … positions), was known to India in the beginning of the Christian era. The earliest recoded evidence of this notation is a copper plate recorded a grant of property in which the date Samvat 346 (595 AD) is given in positional notation. This notation passed on to the Arabs in the eighth–nineth century and from there to Europe in twelfth–thirteenth century. However, the cumbersome Roman numerals were in common usage in Europe till the seventeenth century; these still do survive in the numbering of preliminary pages in a book.

It is worth recalling the words of the famous French mathematician Laplace (1749–1827) in this connection, "It is India, that gave us the ingenious method of expressing all numbers by means of ten symbols, each receiving a value of position as well as an absolute value, a profound and important idea which appears so simple to us now that we ignore its true merit. But its very simplicity, the great ease which it has lent to all computations, puts our arithmetic in the first rank of useful inventions; and we appreciate the grandeur of this achievement the more when we remember that it escaped the genius of Archimedes (Archimedes of Syracuse, 287–242 BC), a leading Greek mathematician, physicist, engineer, inventor, and astronomer of classical antiquity, and Appollonius (Apollonius of Perga, 262–190 BC, known as 'The Great Geometer'), two of the greatest men produced by antiquity."

3.4 1900–2015 AD: impact on zero due to modern digital computer

The advent of *a numerical zero* (as opposed to true zero or, equivalently, the mathematical zero) has been most remarkable. The numerical zero is inseparably connected with the relative error and consequently permits us to know the quality of the solution/output. The cost of producing the solution, however, depends on the specific algorithm (out of many)/computer program used for the solution. *It can be seen that error-free solution of a problem involving a continuous quantity (say, milk) is impossible since the inputs themselves are always in error and are never known exactly.* On a computer, an *error implying a pair of error-bounds that contain the exact error* is a constant companion of a computation involving physical quantities (which are, in general, numerically nonrepresentable in exact numerical form).

3.4.1 Natural, regular, computational mathematics with calculus, and role of zero

We discuss the vital characters, including the consistency and its proof aspects, of natural mathematics which knows no error, that of computational/computer mathematics which is, in general, ever entangled with nonseparable error, and that of mathematics which attempts to capture/approximate the natural mathematics and involves a combination of an erroneous numerical form and an error-free symbolic form. We then provide the scope and importance of understanding the distinction among these three mathematics with stress on scientific and engineering computations which constitute the interface between the mathematical models and their engineering implementations and which pervade every moment of our daily lives.

Morris Kline (1908–1992) stated philosophically "*God exists since mathematics is consistent, and the devil exists since we cannot prove the consistency.*" In the same spirit, we write "*God exists since computer mathematics is consistent, and the devil does not exist since we can prove the consistency.*" and "*God exists since natural mathematics is consistent, and the devil does not exist since the inconsistency is completely unknown in nature.*"

There are three kinds of distinct mathematics: natural mathematics (NMA), computational/computer mathematics (CMA), and regular mathematics (RMA). It is Nature that feeds all the required inputs such as the seed, heat, light, and nutrients, and executes all the functions which outputs a tree. Implicit in these functions is the perfect mathematics used by nature observing all the laws of nature—not all exactly known to us or comprehended by us. It is the absolute fact that there exists no chaos, no violation of any law, and no fuzziness in nature. The low pressure created over Bay of Bengal and consequent cyclone then entering into coastal cities leaving a trail of destruction of properties and lives have never been whimsical. These have followed all laws and underlying mathematics completely errorlessly, perfectly contradiction-freely and absolutely in parallel, and have chosen their paths 100% accurately.

We, the human beings, have very inadequate knowledge and computational power to predict their paths sufficiently accurately. A prediction (partial differential equations/Navier–Stokes equations) model designed by humans using their insufficient knowledge and also tools is no match to those of nature. Consequently, the prediction could be completely or partially wrong. Nevertheless we, the mathematicians and computational mathematicians/physicists, are ever learning and ever improving the model and ever attempting to come as close as possible to those activities and NMA. Our mathematics, physics, and computational mathematics may never come reasonably close to NMA/natural science (NSC—sciences performed by nature), although we will never cease to pursue and improve in our quest of understanding Nature more and more.

RMA or, simply, mathematics is viewed differently by different people. Some of the views are:

- Mathematics is the study of "quantity." (Aristotle, around 384–322 BC)
- Mathematics is the door and key to the sciences. (Roger Bacon, 1214–1294)
- Mathematics is the science of order and measure. (Descartes, 1596–1650)
- Mathematics is concerned only with the enumeration and comparison of relations. (Gauss, 1777–1855)
- Mathematics is the science of what is clear by itself. (Jacobi, 1804–1851)
- Mathematics today is the instrument by which the subtle and new phenomena of nature that we are discovering can be understood and coordinated into a unified whole. (Homi Jehangir Bhabha, 1909–1966, Indian nuclear physicist).
- Mathematics freeze and parch the mind, … an excessive study of mathematics absolutely incapacitates the mind for those intellectual energies which philosophy and life require, … mathematics cannot conduce to logical habits at all, … in mathematics dullness is thus elevated into talent, and talent degraded into incapacity, … and mathematics may distort, but can never rectify, the mind. (Sir William Hamilton, 1788–1856, the famed Scottish philosopher, logician, and meta-physicist)—A negative view that may be construed as a cruel attack on mathematics and hence on mathematicians.

In this context the story "The Blind Men and the Elephant" (from the Buddhist Sutra) is significant. "Why? Because everyone can only see part of the elephant. They are not able to see the whole animal. The same applies to God and to religions. No one will see Him completely." By this story, Lord Buddha teaches that we should respect all other legitimate religions and their beliefs. This famous "blind men" episode is no disrespect to any mathematician/celebrity. It is only to point out every individual's own limitation to view the endless scope of mathematics in all sciences and engineering in its entirety.

3.4.2 Mathematics is all-pervading and springs from purity of mind

Underlying all scientific/engineering discoveries, we almost always find the validation/justification through mathematics. Whether it is physics or chemistry or biology or civil/mechanical/electrical/electronic engineering or architecture, or computer/space science/technology, it is the mathematics that pervades all these subjects in a completely inseparable manner. If one tries to remove mathematics/mathematical thinking out of these subjects, then these subjects collapse entirely. However, it is the essence of physics and numbers (integer/discrete quantities) that form the pillar of the foregoing science and engineering/technological subjects. At the base of physics and numbers lies the relevant mathematics/logical thinking of mathematical nature.

It is the endeavor of mathematics and mathematical sciences to come as close as possible to NMA and NSC so that prediction and decision models are continuously improved. Both NMA and RMA have infinity embedded in them. An infinite series such as the series $1 + (1/2) + (1/4) + (1/8) + (1/16) + \ldots$ has the exact finite value 2. The infinite series viz. $1 + (1/2) + (1/3) + (1/4) + (1/5) + \ldots$, on the other hand, has the value infinity; it is even greater than 10^{100}. One may convince oneself readily using Matlab or possibly even using a pocket calculator. While error is recognized as an important matter in both RMA and CMA for consideration, it is nonexistent in NMA. While *proofs* in mathematics and computational mathematics are essential ingredients for acceptance of statements, such *proofs* in natural mathematics are not so much meaningful to be called for.

One and all, including mathematicians, will accept the outcome of natural mathematics as 100% perfect, 100% errorless, 100% assumption-free, 100% axiomatic, and 100% all-natural law abiding. This foregoing statement can never be proved by the conventional mathematical proof procedures since many concerned axioms in nature are either not known to a mathematician or known rather imperfectly. While assumptions whose main purpose is to introduce error and deviation from true happening/outcome so as to make a reasonable solution possible by a human being using the available tools, are essential ingredients in RMA as well as in CMA, these are unknown in NMA. In fact, NMA simply does not need any assumption.

3.4.2.1 Axioms in nature

The nature or, equivalently, the universe is ever dynamic and is completely based on its own mathematics and its perfect unchangeable eternal absolutely correct rules/laws. Newton's three physical laws of motion are axioms (nondubitable truths for mathematicians) in nature/science. It is not difficult to appreciate the truth by us through our experience/experiment/observation. Thus these laws are assertive sentences which are correct axioms in a macro-level natural activity.

3.4.2.2 Science is the second system of causes and so is mathematics

It may be mentioned that science is the second system of causes and so is mathematics. Consider the question "Why" in the following instances, where these Why's cannot be answered by science (e.g., physics).

- Water formation (Two parts of nascent hydrogen reacts chemically with one part of nascent oxygen to make one part of water (H_2O). Why chemical reaction?)
- Gravitational Pull (Why Pull and not Push?)
- Proof by induction/contradiction (Why should induction and/or contradiction prove the theorem?)

3.4.2.3 Assumption versus axiom and their need in proof

The nature of an axiom and that of an assumption are similar. While an axiom is the perceived truth and does not distort the truth, an assumption introduces errors and does distort the truth.

The mathematical proof of any statement is impossible if we are not allowed to take the help of axioms (so called self-evident truths)/assumptions implicitly or explicitly. Thus the proof remains eternal if the concerned assumptions/axioms remain eternal. "Self-evidence" is not acceptable by a mathematician as a valid proof in mathematics.

We have other different types of proof such as statistical proof, probabilistic proof, documentary proof, evidential proof, and experimental proof. We will not dwell on these proofs as these are not so strictly mathematical proofs nor are these as flawless as mathematical proofs.

3.4.2.4 Computational mathematics

CMA is the interface between mathematics (RMA) and real-world engineering/ technological implementation and beyond, where mathematics has its own limitations There are three types of CMA: *Numerical* (mostly arithmetic operations on numbers), *Semi-numerical* (roughly 50% arithmetic and 50% nonarithmetic operations), and *Nonnumerical* (operations such as testing, branching, looping). Here we confine ourselves to only numerical aspects of computational mathematics connected with modern digital computers.

CMA differs from RMA in that it does not include *infinity*. It, however, includes unique absolute (mathematical) zero as well as *numerical zero* and has a finite range -10^{100} to -10^{-50}, 0, $+10^{-50}$ to $+10^{100}$ (say). A division by the absolute zero in CMA is considered to be a serious violation or, equivalently, a blunder; this is also so in both NMA and RMA, although such division by zero in NMA never occurs. Even division by a numerical zero (a relative zero which is nonunique) is not permitted in CMA (in the context). *Since NMA is essentially out of bound, we focus more on RMA and CMA.*

We compare RMA with CMA followed by a discussion on Ultra-high Speed Computing (UHC) with Dynamic Domain of Applications in brief. We then mention emerging areas/problems in engineering and the effect of the dynamically changing face of computational mathematics to tackle these problems along with exponentially galloping computational speed, storage space, and band-width along with conclusions.

3.4.2.5 RMA versus CMA (Numerical)

The exponential growth that the world has witnessed during the last five decades (1960s onward) in architecture, engineering, and technology, specifically, information technology has become possible due to fantastic progress in both computing devices (hardware) and the software. *Every 18 months processor speed is doubling. Every 12 months band-width is doubling and every 9 months hard disk space is doubling.* Behind all these exponential growth is the computational mathematical algorithms (in the form of software) superimposed on the hardware. Here we can see the real power of RMA through its computational arm. All sciences and all engineering that we see today would have remained dwarfed without this computational arm of mathematics. If we cut this arm, the whole of today's scientific world would simply collapse.

3.4.2.6 Inputs and outputs make a vital difference (infinite versus finite precisions)

Inputs to RMA and that to CMA along with their outputs make a vital difference between them. CMA deals with *only finite precision* and will continue to do so through eternity; there is always a gap in between any two points. RMA, on the other hand, encompasses both finite as well as infinite precisions while NMA eternally works on *only infinite precision*. One may visualize 2-digit finite rational points (denoted here by asterisks on a (2-D) plane) by creating the two-dimensional figure using Matlab commands >>*n = 99;t = 0:1/n:1,x = t; y = round(rand(n+1, 1)*(n+1))/(n+1), size(y), maxi = max(y), mini = min(y), plot(x, y, '*')*. If the precision of the computer is 2 digits, then the foregoing finite (nonnegative) rational numbers produced by the commands could be the inputs in computational mathematics.

There are distinct gaps (not necessarily equal; as a matter of fact, often unequal) in between any two points. These gaps should not be construed as random. On the other hand, in NMA, real numbers (totality of rational and irrational numbers) when represented as blue/black points on a white plane depict no gap between any two points. Hence the plane is entirely filled-in and thus is an absolutely nonempty continuous (nonwhite) blue/black space (plane).

3.4.2.7 A point and a building block of matter: analogous?

It is interesting to distinguish between a physicist's concept of a building block of matter and that of a point representing a real number. A particle physicist has been in the quest of finding an indivisible building block of matter where no further division of this block exists. He has never visualized (or even imagined) matter as something having continuous mass not made of tiny particles or, equivalently building blocks with spaces/gaps between any two blocks/particles.

In the Rutherford-Bohr model of Atomic Structure (1913 AD) as described in Chapter 2, an atom is very stable due to electromagnetic force between negatively charged electrons and the positively charged nucleus. This basic model involved no subatomic elementary particles, each having low or medium or heavy mass.

Electrons, at 297.15 K (room temperature), rotating at different orbits with nucleus at the center of the atom are relatively very far away from the nucleus like our planets (analogous to electrons) revolving round the sun (analogous to nucleus) in the solar system (analogous to the atom). When the temperature is reduced increasingly, the enormous gap between an electron and the nucleus reduces, the kinetic energy of the electron decreases, and the volume of the atom shrinks (all three increasingly). When the temperature is numerically brought down to 0 K, the whole atom reduces to a mass having numerical zero volume and zero energy state as if the whole matter of the system consisting of such atoms is no longer comprised of trillions of extremely tiny building blocks such as atomic/subatomic particles.

It possibly returns to a shapeless and an attributeless phenomenon.

In a devolution, it reverts just to potential energy form to manifest as something not exactly experienced by us. Will a physicist be able to appreciate the concept of point representing a real number (infinity of them and uncountable even in an extremely small space/plane) of RMA/NMA in terms of an analogy of a physical phenomenon? However, the concept of a point (finite in number and countable) in CMA can be easily appreciated by a physicist or for that matter, by anybody in science and engineering and possibly even an analogy of this point could be seen by a scientist/engineer in the physical world.

In RMA, a three-dimensional point is defined as something which has no width, no length, and no height and yet it exists or needs to be assumed to exist for the purpose of all sciences and engineering activities. It is more of a concept that one conceives rather than a definition that appears contradictory. A foregoing building block that a physicist talks about may be somewhat (not completely) analogous to a point. In this respect, a mathematician's view of a real number is not analogous to an elementary particle (since a real number could have infinity of digits).

3.4.2.8 RMA/NMA versus CMA: vital differences

Table 3.1 depicts vital differences between RMA/NMA and CMA.

3.4.2.9 How CMA solves RMA problems easily: example

If there is a break (consisting of more than one point or, equivalently, infinity of points) or a jump discontinuity (as per RMA), the CMA finds out and tells us. As a matter of fact, Matlab does this by finding the computational limit of the function at the given point from both sides (from the left as well as from the right) in a 1-D domain if the function does not exist or undefined mathematically (0/0 form or infinity/infinity form or infinity/0 form) at this given point. In a 2-D domain, that is, for the function f(x, y), we compute using Matlab, four values $f(x \pm \Delta x, y \pm \Delta y)$ to find out the limit. Similarly, we can compute the limit of multidimensional functions.

If an engineer/a nonmathematician, who wants mathematics not as a strict discipline but as a tool for real-world usage, is asked the value of the function f(x) = (x^2 − 4)/(x − 2) at x = 2, his answer will be 4. But a mathematician's answer will be "undefined." It may be remarked that a mathematical solution such as a solution of the Laplace (partial differential) equation $u_{xx} + u_{yy} = 0$ viz. $u(x, y) = e^x \sin y$ is of no use to an engineer unless this is converted to numbers subject, of course, to the appropriate numerical boundary conditions. The domain of problems that can be solved using CMA is much larger than that using RMA. In fact, if a problem is not solvable computationally, it cannot be solved by any other means for the use by an engineer.

3.4.2.10 Godel's incompleteness theorem: blow to complete nonfuzziness of mathematics

David Hilbert (1862–1943), a great German mathematician, proposed, at the beginning of twentieth century, 23 problems which, he believed, needed to be solved in all parts (of Hilbert's program) to put solid logical foundation under all of mathematics. Kurt Godel (1906–1978), a brilliant Austrian logician, showed in a proof that any part of mathematics at least as complex as arithmetic can never be complete. No algorithm, howsoever large, can lead to sorting out all the true or untrue statements/information/equations within a system. He demonstrated that statements exist that cannot be derived by the rules of arithmetic proof.

A simple explanation of Godel's Theorem A statement P which states that "there is no proof of P." If P is true then there is no proof of it. If P is false then there is a proof that P is true—which is a contradiction. Therefore, it cannot be determined within the system whether P is true.

Table 3.1 Mathematics/Natural Mathematics (RMA/NMA) versus computational mathematics (CMA): Vital differences

Mathematics (RMA/NMA)	Computational mathematics (CMA)
Domain has ∞ (infinity) of points.	Domain has finite number f of points.
2-D. domain has ∞^2 of points each represented by two ordered mathematically real numbers.	2-D domain has f^2 of points each represented by two ordered rational numbers, each with finite number of digits.
Proof methods implicitly assume n-D domain having ∞^n points.	Proof methods may simply check validity for each point of n-D domain with f^n points.
Methods of induction/deduction/contradiction/construction form the backbone of mathematical proof. NMA needs no such methods for proof and even no proof; all its statements are absolutely error-free perfect axioms.	These methods are not essential. Exhaustive verification provides the proof. In UHC, such a verification for a specified precision is always possible unless it is not too intractable (e.g., Chess problem).
Assumptions only introduce mathematical error and deviation from the true happenings. NMA is completely assumption-free while RMA is not.	Both assumptions and numerical computations introduce error and deviation more than what RMA projects.
Mathematics in Natural Computer (NMA) is 100% errorless, assumptionless, and only axiom-oriented. RMA encompasses errors, assumptions, and also axioms.	Mathematics in a man-made computer (CMA) is intrinsically erroneous and includes more assumptions than RMA besides axioms.
Natural Computer is the most parallel and fastest computer in the universe—started infinite years ago, working now, and will continue to work infinite years hence.	Artificial modern computer has relatively very little parallelism, and too slow. It started about seven decades back (1940s) and will continue for a finite time as long as civilization lives.
Natural Computer knows no break-down and hence no maintenance.	Artificial computer knows occasional break-down and hence preventive and break-down maintenances.
Natural Computer needs no resources/help from humans/any living being. It operates 100% independently and is never stoppable.	Man-made computer needs resources and help from humans. It operates depending on availability of resources (e.g., electricity) and humans and is ever stoppable.

3.4.3 *Ultra-high speed computing with dynamic domain of applications*

The fastest computer chip, announced by IBM, in 2011, that integrates both electrical and optical nanodevices on the same piece of silicon could soon make it possible for UHC to execute 10^18 (i.e., one million trillion) flops (floating-point operations per second). This chip is based on IBM's CMOS Integrated Silicon Nanophotonics (CISN) technology at Semicon, Japan.

The information processing of the brain is extremely fast, highly parallel, and much less mistake-prone in certain situations such as the situation where one is to recognize a given picture of her mother which she does within a fraction of a second. A Cray super-computer, on the other hand, would take several seconds to come out with an answer that the picture is that of her mother. In other situations, however, the information processing by the brain of a common human being could be slow and could involve error/mistake.

In a conscious state of mind, the processing is often sequential (nonparallel), slow, and could involve error while in a subconscious/unconscious state of mind, the processing could be parallel, fast, and less error-prone. There exists no living being who/which could claim that she/it could process information all the time error-free. As a matter of fact "To err (mistake) is human (living being) while not to err is computer (nonliving being)" could be an extension of the age-old proverb "To err is human."

So far as nature/material universe is concerned, it is not difficult to appreciate that perfect mathematical activities involving exact real numbers is continuously going on with the highest possible parallelism and in the fastest possible manner. These mathematical activities of the universal/natural computer is never stoppable and is completely error-free, totally maintenance-free, absolutely assumption-free, as well as entirely bug-free. The concerned natural mathematics was existing infinite years ago, it is existing now, and it will continue to exist infinite years hence and follows all the laws of nature all the time perfectly.

The digital computer is based on silicon technology. That is the only technology which alone has revolutionized the whole world since mid-twentieth century. Both the technological and architectural innovations in silicon technology over decades are responsible for dynamic and exponential increase of processing power, band-width, and storage space (executable memory and hard-disk space). The power is further enhanced significantly by the use of several central processing units as well as other processors/channels/input-output controllers.

Moreover, unlike the mainframe days when many users would use one large computer in a time-sharing mode, we have numerous (hundreds of millions) personal computers, both desk-tops and laptops, with a huge processing power available to the world. Only a small fraction of this power is utilized. The remaining huge processing power unutilized is a huge waste. The life of a computer is not reduced by using this untapped power, nor the cost of processing is increased by using all of the power—a situation very much unlike our daily consumption of electricity, water, gas, and other necessities. Consequently, many computing problems—such as NP-hard problems—that were intractable have now become tractable within a reasonable precision and a reasonable size with considerable practical importance.

Evolutionary, as well as trial and error, approaches including genetic algorithms are being more increasingly employed for highly compute-intensive problems with relative error computation to ascertain the quality of the solution. Even in many situations such as the computation of an integral involving complex trigonometric and special functions, such an approach produce better quality of the integration value than that produced by a good quadrature formula. The exhaustive search over a reasonably large finite practical domain is tractable. Consequently the validity of any statement (lemma/theorem/corollary) can be checked in CMA; no mathematical proof methods such as those of induction, contradiction, deduction, construction and a combination of two or more of them are required.

Cloud/grid computing and other form of future computing are now aiming at reducing/minimizing huge unutilized computing power of numerous computers available.

3.4.4 Impact of emerging engineering on RMA, CMA, UHC, and their teaching

Not only the area of computing/computational sciences is fast changing but also the area of RMA as well as those of engineering are rapidly changing. All these areas are interdependent and all are progressing by leaps and bounds. Such a progress has been possible due to exponentially increasing computing speed, storage, and band-width with practically 100% reliability.

3.4.4.1 RMA, CMA, UHC are interconnected: domain knowledge and polynomial-time algorithms are desired

Stand-alone mathematics (RMA) is unthinkable today. Also doing RMA for academic interest only has highly limited scope. Further, each of CMA and supercomputing or, equivalently, UHC or hyper-computing cannot be stand-alone. Both are closely related so far as their real-world implementations are concerned. RMA, CMA, and UHC have to be implicitly or explicitly embedded in all sciences and engineering/technology and readily useful to the society. For this we need to have a fairly good view of the trends that are emerging in engineering during the coming decade(s). These three areas as well as their teaching need to be changed/oriented based on the foreseeable technology/engineering activities.

Consequently the domain knowledge, that is, knowledge of the concerned technology/engineering activities, has to be acquired by the scientists involved in RMA, CMA, and UHC. All the emerging engineering activities have one thing in common. This is the demand on more and more computing power, that is, more and more computing resources such as processing speed, band-width, and storage space (both main memory and hard disk) since these activities are becoming increasingly highly computer-, storage-, and communication-intensive.

Besides, there is always a need for more efficient polynomial-time algorithms with optimized storage, intra-algorithm, and inter-algorithm communications. Wherever such a polynomial-time algorithm is currently unavailable or such an algorithm is exponential/combinatorial the concerned scientists need to attempt to devise/design a polynomial-time algorithm mathematically. Also, they should devise best possible evolutionary approaches whose computational and storage complexities are polynomial-time. Here we first describe the trends in engineering/technology during the coming decade and then based on these trends we put forth how mathematics teachers should orient themselves to adopt the dynamic situation in applicable science and technology.

3.4.4.2 Emerging trends

The current technology/science trends are, among others, space tourism/travel, nanotechnology, mapping of the human genomes, natural (e.g., human/living being) brain research, social media, face transplant, hybrid cars, global communications, nonsilicon technology such as quantum computing technology, digital storage, subatomic/sub-subatomic particles as building blocks of the universe, nonmaterial (spiritual) science pervading and injecting life into matters, cancer cures, synthetic biology, biointerfaces, long time storage of solar, wind, and wave energy, and smart grids. Since conventional materials being used currently do not work sufficiently well, radical materials (products based on carbon nanotubes) are becoming increasingly promising.

3.4.4.3 RMA, CMA, UHC, and their teaching should be oriented based on dynamic domain knowledge/requirements

Some of the other areas, besides the aforementioned ones, which are very important for the society and are emerging in a big way are data interface and managing global warming. These emerging areas necessitate RMA, CMA, and UHC to orient themselves according to their dynamic needs.

It may be observed that a mathematical solution is of no use to an engineer until it is translated into numbers since he will be able to implement numerical solution for building bridges/machines/mansions but not the mathematical (symbolic) solution. Any mathematics teaching should include CMA teaching. It is particularly so because UHCs are available individually with practically everybody. Mathematics teaching without CMA and UHC will neither be enjoyable nor be satisfactory to most of the students.

Lots of innovative activities as well as dynamic teaching (to produce sufficiently able human resources) in RMA, CMA, and UHC are needed. Both the deterministic and evolutionary algorithms best suited for the concerned physical problems need to be designed and developed in the fast moving scenes in the years to come. The society and the world will then benefit highly.

RMA, CMA, and UHC (combined together) give an enormous amount of insight into the physical problem posed and hence implicitly help the physical scientists to appropriately modify the problem formulation and then resolve in a comparatively short tolerable span of time. The new solution will expose us to the problem still further and thus recursively help us to improve the problem formulation and the successive solutions dynamically. It is necessary to stress that the current usage of a computer to solve a problem/successive modified problems is not a one-time affair as was used to be practically so a few decades ago. This is because every individual has a high-speed computer all to herself unlike a single mainframe (centralized) computer available to many in a time-sharing mode during the 1960s, 1970s, and even 1980s.

3.4.4.4 How zero is connected with foregoing subsections: quality of a numerical zero

Mathematics (RMA) is viewed differently by different people over centuries. In essence, it is all-pervading and springs from the purity of mind. There are three gigantically different mathematics, that is, NMA, RMA, and CMA. While NMA is completely error-free, ever-existing, nonstoppable, and mostly beyond the comprehension of human beings, RMA and CMA both have an error component, are consciously stoppable, and are within the comprehension of human being.

RMA has been continuing to emulate NMA as best as possible. CMA uses UHC and the knowledge of RMA to manifest how good/close the solution/result is compared with the actual physical outcome. Twenty-first century RMA and its computational arm (CMA) along with UHC have impacted immensely the ongoing engineering and technological activities (ETA). All of these, that is, RMA, CMA, UHC, and ETA are mutually impacting one another and are enriching themselves fast dynamically through nonstop innovations. Not only the current ETA but also the emerging ETA have affected profoundly both RMA and CMA.

The exponentially galloping computing power with *increased precision* (and also storage) has helped in achieving all that we see in the real world today, which possibly was beyond our imagination in the 1950s. Teaching RMA, CMA, and UHC along with that of the concerned domain (physics, chemistry, biology, engineering, or art) is also impacted and will (will have to) continue to remain dynamic to meet the future need of trained and innovative manpower in all spheres of human activities. The central issue is the required quality of the solution which implies required narrow error bounds that depend on the concerned numerical zero in the computation. Increased precision allows improved numerical zero, that is, a zero closer to exact zero.

3.4.5 Exponential growth of computing power has made all achievements up to 1950s dwarf

The exponential growth that the world has seen during the last five decades (1960s onward) in architecture, engineering, and technology, and in particular, information technology has become possible due to fantastic progress in both computing devices (hardware) and the software (including firmware). *Every 18 months processor speed is doubling. Every 12 months band-width is doubling and every 9 months hard disk space is doubling.* Behind all these exponential growth is the computational mathematics superimposed on the hardware. All the foregoing views—both negative and positive—on mathematics need to be modified when we land into the world of computational mathematics. Here we can see the real power of mathematics through its computational nature. *All sciences and all engineering that we see today would have remained dwarfed without this computational arm of mathematics.* At the root of all sciences and technology is this arm. If we cut this arm, the whole of today's scientific world would simply collapse.

3.4.5.1 Inputs and outputs make a vital difference

Inputs to RMA and that to CMA along with their outputs make a vital difference between RMA and CMA. Depiction of 2-digit finite rational numbers and real numbers as random points are displayed in Figures 3.1 and 3.2 respectively.

3.4.6 Floating-point representation of numbers and arithmetic

The floating-point representation of numbers corresponds closely to "scientific notation"; each number is represented as the product of a number with a radix point and an integral power of the radix. One bit of the word is for the sign of mantissa, e bits of the word for the exponent while f bits for the mantissa or, equivalently, significand as in Figure 3.3.

The exponent bits (usually in excess 2^{e-1} code) represent the actual integer E. The fraction (mantissa) bits represent the fraction F, where $0 < F < 1$. The number in the computer word would be $\pm F \times 2^E$. In other schemes, the value is taken to be $\pm F \times B^E$ for some constant B other than 2. IBM 360/370 computers use $B = 16$. Here we will consider $B = 2$. The exponent may be positive or negative. The sign bit represents the sign of the mantissa. The exponent expressed in excess 2^{e-1} code takes care of the sign of the exponent. If all the e bits are 0 then these bits will represent the actual exponent $-2^{e-1} = -128$ when the number of bits $e = 8$, that is, the actual multiplier is $2^{-128} \approx 0.350 \times 10^{-45}$. If the leftmost bit (most significant) bit of e bits is 1 and the rest are zero then these bits will represent the true exponent $2^{e-1} - 128 = 0$ when the number of bits $e = 8$.

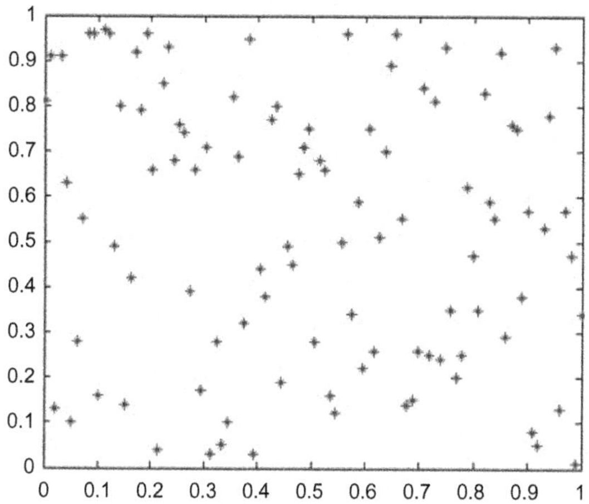

Figure 3.1 Depiction of 2-digit finite rational numbers as random points denoted here by asterisks on a (2-D) plane. If the precision of the computer is 2 digits then the foregoing finite (nonnegative) rational numbers could be the inputs in computational mathematics. There are random gaps in between any two points. The figure is created by the Matlab commands *>>n = 99; t= 0:1/n:1, x = t; y = round(rand(n+ 1, 1)*(n + 1))/(n + 1), size(y), maxi = max(y), mini = min(y), plot(x, y, '*').*

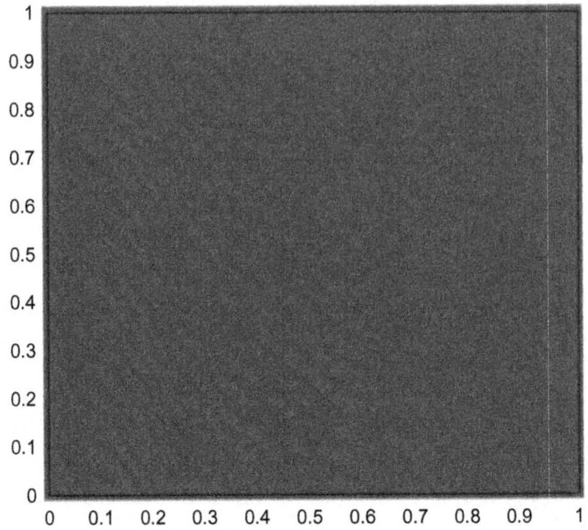

Figure 3.2 Depiction of real numbers (totality of rational and irrational numbers) as random points on a (2-D) plane having no gap in between any two points; hence the plane is entirely filled-in and thus is a continuous blue/black space.

Figure 3.3 Floating-point number format.

3.4.6.1 Dwarf and machine epsilon

In a 32 bit word, if one bit is for sign, 8 bits for exponent, and 23 bits for mantissa then the concept of the dwarf and the machine epsilon is as follows. The smallest representable number which is just greater than 0 in magnitude is called the dwarf (Figure 3.4). It would be, allowing **0** (bold) = 0000,

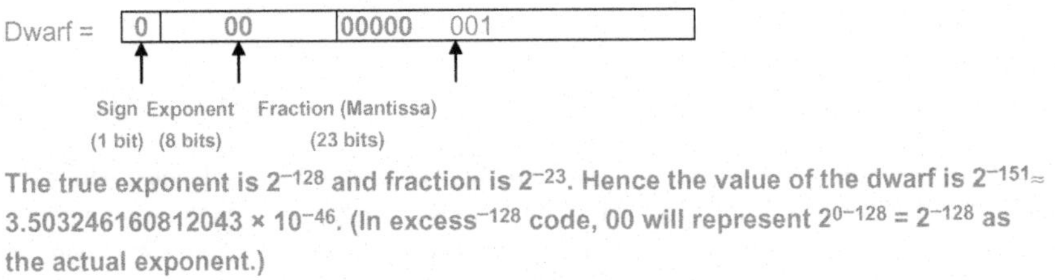

Dwarf =
0
00
00000 001
Sign Exponent Fraction (Mantissa)
(1 bit) (8 bits) (23 bits)
The true exponent is 2^{-128} and fraction is 2^{-23}. Hence the value of the dwarf is $2^{-151} \approx$ $3.503246160812043 \times 10^{-46}$. (In excess^{-128} code, 00 will represent $2^{0-128} = 2^{-128}$ as the actual exponent.)

Figure 3.4 The dwarf (smallest representable floating-point number just greater than 0) in a 32-bit word with 23 bit mantissa and 8 bit exponent.

The machine epsilon (Figure 3.5) is the smallest number that a computer recognizes as being very much bigger than zero as well as the dwarf in magnitude. This number varies from one computer to another. *Any number below the machine epsilon, when added or subtracted to another number, will not change the second number.* It is represented, allowing **0** (bold) = 0000 (a block of four bits), as

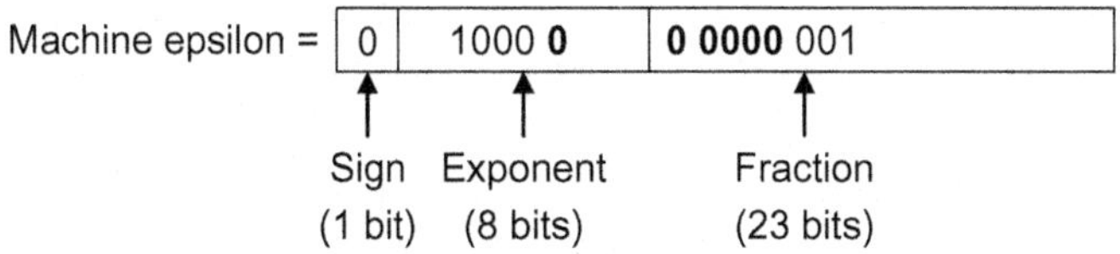

True exponent = 2^0 (in excess-128 code), machine epsilon is $2^{-23} = 1.19 \times 10^{-7}$.
(In excess-128 code the exponent is in [2^{-128} to 2^{127}]).

Figure 3.5 Machine epsilon (the smallest number recognized by the computer as very much greater than zero as well as dwarf in magnitude and when added to 1 produces a different number).

During computation if a numerical value $\alpha <$ machine epsilon (and, of course, larger than the dwarf) then adding this value α to another value b will keep the result b only. Such a situation may lead to an *infinite loop* if b is tested against another value c. A legal way for jumping or breaking out of the loop is shown in the following Matlab program called *mcepsilon*:

%mcepsilon

```
eps=1; format long;
  for n=1:500
    eps=eps/2; if (1+eps) <=1, eps=eps*2; break; end
  end
  n, eps
```

The machine epsilon (in double precision) is *eps = 2.220446049250313e−016*. It is obtained when the number of terms is n = 53. This Matlab program shows a way of estimating the machine epsilon of a specified machine (here in double precision), in which the *for loop* is executed for sufficiently large number of times.

Using the Matlab hexadecimal format, viz., format hex instead of the decimal format, viz., format long, we would be able to see that the machine epsilon when added to any value in between ≥1 and <2 will change the value. But when it is added to any value ≥2 will not change the value. For example, after running the foregoing mcepsilon (Matlab) program, we obtain

```
n = 53, eps = 2.220446049250313e-016.
```

Now using the following Matlab commands

```
» format hex
» eps
```

we obtain *eps=3cb0000000000000*. The following Matlab command

```
» 1+eps
```

gives the result *ans=3ff0000000000001* whereas the representation of 1 in hexadecimal format is *3ff0000000000000* which is different in the last hexadecimal digit. The command

```
» 1.9999999999999+eps
```

produces *ans = 3fffffffffffe3f*. The command

```
» 1.9999999999999
```

gives *ans = 3fffffffffffe3e* which differs in the last (least significant) hexadecimal digit. The commands

```
» 2+eps and » 2
```

produce the same result *ans = 4000000000000000*. Also, the commands

```
» 3+eps and » 3
```

produce the same result *ans = 4008000000000000*

The Matlab program called dwarf for determining the dwarf may be written as follows.

```
%dwarf
eps=1; format long;
  for n=1:1500
    dwarf=eps;
    eps=eps/2;
    if eps==0, break; end;
  end
  n, dwarf
```

The value of dwarf (in double precision) is given as *dwarf = 4.9406564584124654e–324* and the corresponding number of terms *n = 1075*.

The floating-point representation provides much larger range of values to be represented unlike the fixed-point representation. The disadvantage of the floating-point notation is that we do not obtain as many as k– 1 significant bits in one word for a k-bit word computer.

If a number less than machine epsilon and obviously greater than or equal to dwarf is added or subtracted to a number k (greater than or equal to machine epsilon), then the result remains k only. Thus repeated such additions/subtractions could produce an infinite loop. *Thus any number less than machine epsilon and greater than or equal to dwarf may be called a local zero with respect to the given computational device.*

A numerical zero in computational science is nothing but a local zero which could vary from one computer to another as well as from one context to another. In other words, it is relative and is not unique. In a modern digital computer, we have a finite number of numerical zeros besides, however, the true (mathematical) zero which is unique and absolute.

3.4.6.2 The true zero, a local/numerical zero, and error

The exact real quantity Q in terms of volume as well as in terms of weight is continuous and is never known. *It was not known in the past. It is not known now. It will never be known in future.* We cannot say with 100% confidence that one litre of milk that we buy from a shop is exactly one litre. What we can say with 100% confidence that the exact quantity lies in (one liter minus α milliliter, one liter plus α milliliter), where α depends on the accuracy of the concerned measuring device and should be as small as reliably measured by the device. On the other hand, if we know the (any) exact quantity Q representable in a finite number of decimal digits in the conventional decimal system, then there is no need to bring error into picture.

While error is never desirable, it has become an inseparable part of every step of our lives. In this context, a local zero or, equivalently, a relative or a numerical zero is in place and is of practical importance (to know the quality of a solution). This local/relative/numerical zero is evidently not unique. It corresponds to any quantity Q′ such that the relative error $E_r = |Q-Q'|/|Q| \leq 0.5 \times 10^{-k}$, where k = 4 (say), the number of significant digits desired. Consequently *a numerical zero* is any quantity $\leq 0.5 \times 10^{-k}$ such that $E_r \leq 0.5 \times 10^{-k}$. The absolute error of a quantity is defined as follows. If Q is the exact quantity and Q′ is an approximate quantity then $E_a = |Q-Q'|$ is defined as the mathematical absolute error while the mathematical relative error E_r is defined as $E_a/|Q|$. The modulus symbol (absolute value) of the quantity is used just to stress on the fact that we do not know whether the error is negative, 0, or positive.

Since the exact quantity Q is never known, we take Q as the quantity of (sufficiently) higher order accuracy and Q′ as the quantity of lower order accuracy. We then can readily compute the numerical relative error or simply relative error in Q′ which is practically all important unlike numerical absolute error or simply the absolute error. *Hence a numerical zero (and thus the relative error) in the context of numerical computation is all-important in practice.* Why is relative error all-important? Is $5 million a big sum or a small sum? It is a very large sum compared to the annual income of an average American while it is a numerical zero compared to the Federal budget of the United States. Thus an absolute error does not provide the most desired information, that is, how important the sum is, to the reader. On the contrary, a relative error or, equivalently, a percentage error, that is, 100 times the relative error, does.

3.4.6.3 Newton scheme for nonlinear equations with Matlab

Consider, for example, the second order Newton fixed-point iterative scheme to compute a root of the real polynomial equation f(x) = 0. The scheme is as follows.

$$x_{i+1} = x_i - (f(x_i)/f'(x_i)),\ i = 0,1,2,\ldots,\ \text{till}\ |x_{i+1} - x_i|/|x_{i+1}| \leq 0.5 \times 10^{-4} \quad (3.1)$$

where x_0 is an initial approximation provided by the user and if four significant digit accuracy is desired. Since no measuring device can usually measure a quantity with an accuracy greater than (or, equivalently, with an error less than) four significant digits, the foregoing scheme will serve the purpose most of the time and particularly it will serve the purpose always if the output of the scheme is not an input to an intermediate step of the physical problem. Observe that the term "significant digit" implies the relative error while the term "decimal digit" implies the absolute error. Further it can be seen that the scheme always converges for any nontrivial real polynomial equation f(x) = 0 and for any finite initial approximation. To obtain a complex root of the real polynomial, it is necessary to choose the initial approximation as a nontrivial complex number. Even if the equation has no complex root, a complex initial approximation will converge to a real root. If the initial approximation is far away from a root, then the number of iterations will be large. It is preferable to choose an initial approximation as close as the root of the equation possible. As a numerical example consider the real second degree polynomial equation $f(x) = x^2-9$. Using the Newton scheme and choosing the initial approximation as $x_0 = 9$ (not a very good one for a square-rooting operation), we obtain through the following Matlab program the output

```
function []=newton70(f, x); format long g;
fp=diff(f);k=0; x1=x-(eval(f)/eval(fp));
while abs(x1-x)/abs(x1)>=.5*10^-4,
  x=x1;fvalue=eval(f);C(k+1,:)=[k+1 x fvalue];
  x1=x-(eval(f)/eval(fp));k=k+1;end;
disp('       iteration       x        f(x)'),C
%To execute newton70, enter into Matlab command line
%clear all;syms x;double x;newton70(x^2-9, 9);
>> clear all;syms x;double x;newton70(x^2-9, 9);
```

The output:

```
      iteration    x                    f(x)
C =
      1            5                    16
      2            3.4                   2.56
      3            3.02352941176471     0.141730103806228
      4            3.00009155413138     0.000549333170440036
```

The root x = *3.00009155413138* obtained at the fourth iteration is correct up to four significant digits. Here the local (relative) zero is any value less than or equal to 0.5×10^{-4}. This is also called the numerical zero in the context. For real world implementation, the root is good enough. It may be remarked that the value of the function $f(x) = (x^2 - 16)/((x - 4)$ at x = 4 will be 8 for an engineer and undefined for a mathematician. While the universal zero is unique and almost always used by a mathematician, the local zero is nonunique, context dependent, and almost always used by a computational scientist/mathematician.

3.4.7 Calculus: ultimate step in understanding mathematical zero since third millennium BC

The next great mathematician to use zero was *Rene Descartes* (1596–1650 AD), the founder of the Cartesian coordinate system. As anyone who has had to graph a triangle or an ellipse knows Descartes' origin is (0, 0). Although zero was now becoming more common, the developers—ancient and modern ones—of calculus (mathematical study of the rates of change at which a quantity changes) would make the final step in understanding zero.

Adding, subtracting, and multiplying by zero are relatively simple operations. But division by zero has confused even super-minds. How many times does zero go into 10? Or, how many nonexistent mangoes go into three mangoes? The answer is indeterminate, but working with this concept is the key to calculus. For example, when one drives to the store, the speed of the car is never constant—stoplights, traffic jams, and different speed limits all cause the car to speed up or slow down. But how would one find the speed of the car at one particular instant? This is where zero and calculus enter into the picture.

If you wanted to know your speed at a particular instant, you would have to measure the change in speed that occurs over a set period of time. By making that set period smaller and smaller, you could reasonably estimate the speed at that instant. In effect, as you make the change in time approach zero, the ratio of the change in speed to the change in time becomes similar to some number over zero—the same problem that is believed (by some) to have stumped Brahmagupta.

By working with *numbers as they approach zero, calculus was born,* without which we would not have NSCs, engineering, and many aspects of economics and finance. In the *Sulvasutras* and the work of *Aryabhatta,* we find curious approximations of π, that is, the area of a circle of radius 1. *Antiphon* (Antiphon the Sophist was also a capable mathematician of late fifth century BCE) and *Bryson* (Bryson of Heraclea, late fifth-century BCE, was an ancient Greek mathematician and sophist who contributed to solving the problem of squaring the circle and calculating pi) tried to find the area of a circle by inscribing it by a regular polygon with an infinite number of sides.

Around 240 BC, *Archimedes* used *Aristotle*'s idea of the potential infinity to generalize the method of exhaustion of *Eudoxus*, which formalized the ideas of Antiphon and Bryson. He used this method to find areas bounded by parabolas and spirals, the volumes of cylinders, areas of segments of spheres, and especially to approximate π by bounding its value between 22/7 and 223/71. Although Archimedes' geometric method (the circumference of a circle lies between the perimeters of the inscribed and circumscribed regular polygons of n sides, and as n increases, the deviation of the circumference from the two perimeters becomes smaller) to approximate the value of π is heuristic (not regarded as final and strict but merely as provisional and plausible), it foreshadowed the *concept of the limit*.

Archimedes' method was later refined for the purpose of making better approximations to π by *Liu Hui, Zu Chongzhi, van Ceulen*, and others. Zu Chongzhi also established a geometric method for finding the volume of a sphere that was similar to Archimedes' and would later be called *Cavalieri's principle*. Geometric and analytic methods for several other quadrature problems were employed by *Pappus of Alexandria, Mahavira, Galileo, Kepler, Descartes, Cavalieri, Fermat, Gilles Personnier de Roberval* (1602–1675 AD), *Torricelli, Wallis, Brouncker, Pascal, Huygens, Barrow, Gregory*, and others. The quadrature problem eventually gave rise to *integral calculus*, which is the assimilation of the geometric and analytic methods and the understanding of the calculus of infinitesimals or *infinitesimal calculus*.

This final step was taken independently by both *Newton* and *Leibniz*. The term "integral" appeared for the first time in a paper produced by *Jacob Bernoulli* in 1690 AD, and the term "integral calculus" was introduced by Leibniz and *Johann Bernoulli* in 1698 AD. Since this pioneering work by *Riemann* and *Lebesgue*, integral calculus has become a universal method for calculating area, volume, arc length, center of mass, work, and pressure. In mathematics, the word quadrature has survived and it refers to only numerical integration. The word *tangent* comes from the Latin "tangere," to touch. The term tangent line is due to Leibniz, 1692 AD, 1694 AD, who defined it as the line through a pair of *infinitely close points* on the curve. In 1629, *Fermat* innovated a general geometric procedure for drawing tangent lines, or simply, tangents on curves whose analytic forms were known. He made this great discovery while trying to find the maxima and minima of these curves.

During 1664–1666 AD, *Barrow* developed a new geometric method for determining tangents on curves. In fact, in the seventeenth century, the tangent line problem became one of the central questions in geometry. Descartes remarked "And I dare say that this is not only the most useful and most general problem in geometry that I know, but even that I have ever desired to know." In Europe, the tangent problem prefigured *differential calculus*, another term that was coined by Leibniz in 1684, which has been credited solely to Newton and Leibniz. For Newton, the calculus was geometrical, while Leibniz took it toward analysis.

However, the *concept of a derivative* had been developed almost 1200 years before Newton and Leibniz by *Bhaskara II*, who provided differentiation of the trigonometric functions, for example (in modern notation), he established $\delta(\sin x) = \cos x \cdot \delta x$. Bhaskara II also gave a statement of *Rolle's theorem*, concluded that the derivative vanishes at a maxima, and introduced the concept of the instantaneous motion of a planet in his collection *Siddhanta Siromani*. Then, in the fourteenth century, *Madhava* of Sangamagramma invented the ideas underlying the infinite series expansions of functions, power series, the trigonometric series of sine, cosine, tangent, and arctangent (these series have been credited to Gregory, Taylor, and Newton), rational approximations of infinite series, tests of convergence of infinite series, the estimate of an error term, and early forms of differentiation and integration.

Madhava fully understood the limit nature of the infinite series. This step has been called the "decisive factor onward from the finite procedures of ancient mathematics to treat their *limit–passage to infinity*," which is in fact the kernel of modern classical analysis. *Parameshvara Namboodri* (around 1370–1460), a disciple of Madhava, stated an early version of the *Mean value theorem* in his *Lilavathi Bhasya*. This is considered to be one of the most important results in differential calculus, and was later essential in proving the *fundamental theorem of calculus*, which shows the inverse character of tangent and area problems.

Torricelli was the first to understand the fundamental theorem of calculus geometrically, and this was extended by *Gregory* (James Gregory, 1638–1675) while *Barrow* (Isaac Barrow, 1630–1677) established a more generalized version, and finally Newton completed the mathematical theory. *Nilakanthan Somayaji* (around 1444–1544), following the footsteps of Madhava and his father Parameshvara, provided a derivation and proof of the arctangent trigonometric series and gave the relationship between the power series of π and arctangent, namely,

$$\pi/4 = 1 - (1/3) + (1/5) - (1/7) + (1/9) - (1/11) + \cdots,$$

which in the literature has been credited to Gregory and Leibniz. *L'Hôpital* (Guillaume Francois Antoine de L'Hôpital, 1661–1704) is known as the author of the world's first text book on differential calculus, but *Jyesthadevan* (around 1500–1600) wrote the calculus text *Yuktibhasa* in Malayalam (a regional language of the Indian state of Kerala) almost 150 years earlier.

Calculus in Leibniz's sense (which became widely used partially because of his carefully selected notation) was extended by Leibniz himself, the *Bernoulli brothers, Euler, Bolzano, Cauchy, Weierstrass, Riemann*, and several other eighteenth and nineteenth century (AD) mathematicians. Their work made calculus one of the most powerful, supple, and practical tools of mathematics. In fact, calculus finds applications in every branch of the actuarial science, business, computer science, demography, economics, engineering, medicine, the physical sciences, statistics, and this list continues growing.

Acknowledging the importance of calculus, *von Neumann* said that "[t]he calculus was the first achievement of modern mathematics and it is difficult to overestimate its importance. I think it defines more unequivocally than anything else the inception of modern mathematics, and the system of mathematical analysis, which is its logical development, still constitutes the greatest technical advance in exact thinking." Today calculus, its advances and abstractions, and its applied branches such as differential equations, optimization, etc. have become a major part of mathematical education at all levels.

In the twenty-first century, zero is so familiar that to talk about it seems like much ado about nothing. But it is precisely understanding and working with this nothing that has allowed civilization to progress. The development of zero across continents, centuries, and minds has made it one of the greatest accomplishments of human society. Because mathematics is a global language, and calculus its crowning achievement, zero exists and is used everywhere. But, like its function as a symbol and a concept meant to denote absence, zero may still seem like nothing at all. Yet, recall the *fears over Y2K and zero* no longer seems like a tale told by an idiot.

3.4.8 Ramanujan and zero with its eternal spiritual significance

Out of several aspects of the Indian mathematician Srinivasa Ramanujan's (1887–1920) life, the following events are extremely spiritually significant. Just knowing that the Indian mathematician Srinivasa Ramanujan had worked in number theory as many others did is too insufficient to appreciate the creation and the extraordinary depth of the insight (unusual farsightedness) of this great genius. His whole being is a manifestation of divinity in terms of numbers. He correlated each number with an entity in the infinite spiritual universe with full import and realization of its significance (and not just a dry superficial import of common thinking of a man). Each mathematical/computational prodigy depicted his genius either in some specific kind of extraordinary fast computation, memorizing huge numbers for a significant length of time, and/or in some kind of specific mathematical character of numbers far ahead of his time (with a vision beyond any common human being). In fact, all prodigies are usually distinct in their extraordinary talents, that is, no two prodigies are exactly identical in their capabilities.

Out of several aspects of Ramanujan's life, the following events are extremely spiritually significant. These seem to differ from other humans with extraordinary computing and other capabilities and may be depicted through the well-known and well-researched activities of modern scientists.

In one moon-lit night Ramanujan and one of his friends walked six miles to go to Nachiarkovil, a town with a site of a Vishnu temple, to witness a religious festival. All the while, Ramanujan recited passages from the Vedas and ancient Sanskrit tomes (Sanskrit scholarly books), and gave running commentaries on their meaning. On another occasion at the age of 21, he went to the house of a teacher, got drawn into conversation, and soon was expatiating on the ties he *saw between God, zero, and infinity*—keeping everyone spellbound till 2:00 in the morning. "... *Losing himself in philosophical and mystical monologues, he'd make bizarre, fanciful leaps of the Imagination that his friends did not understand but found fascinating anyway. So absorbed would they become that later all they could recall was the penetrating set of his eyes.*

... Ramanujan never did rebel. He did not deny the unseen realm of spirit, nor even held it at arm's length; rather, he embraced it. His was not a life set in tension with the South India from which he came, but rather one resonating to its rhythms."

Godfrey Harold Hardy (1877–1947), an English mathematician, known for his achievements in number theory and mathematical analysis, who was instrumental in inviting Ramanujan to England, was a confirmed atheist. Yet when Hardy died, one mourner spoke of his profound conviction that *"the truths of mathematics described a bright and clear universe, exquisite and beautiful in its structure, in comparison with which the physical world was turbid and confused. It was this which made his friends ... think that in his attitude to mathematics there was something which, being essentially spiritual, was near to religion."*

The same, but more emphatically, is true for Ramanujan. He had profound faith in God and built the landscape of the Infinite, in the realm of both mathematical and spiritual, his home. *"An equation for me has no meaning,"* he once said, *"unless it expresses a thought of God."*

> *...One idea Ramanujan bruited about dealt with the quantity $2^n - 1$. That, a friend remembered him explaining, stood for "the primordial God and several divinities. When n is zero the expression denotes zero, there is nothing; when n is 1 the expression denotes unity, the Infinite God. When n is 2, the expression denotes Trinity; when n is 3, the expression denotes 7, the Saptha Rishis (seven spiritual scientists), and so on."*

Ramanujan was always congenial to metaphysical aspects. The following incident that occurred in Kumbakonam, TamilNadu is an example of his extraordinary behavior toward the gymnastics instructor, Satyapriya Rao, whose excited outpourings which were considered even by many tolerant South Indians as signs of mental imbalance. *"He would stand there, by the Cauvery, staring into the sun, raving; sometimes he'd have to be chained up when he got too hysterical. Most people ignored him. But not Ramanujan, who would sometimes collect food for him; some thought he must be mad to indulge him so. Yes, Ramanujan explained, he knew the man had visions, saw tiny creatures. But in an earlier birth, he was sure, Satyapriya had earned great merit. What others wrote off as the ravings of a madman was actually a highly evolved vision of the cosmos."*

Ramanujan, later, in England, built a theory of reality around Zero and Infinity, though people around him could not quite figure out what he was getting at. *"Zero, it seemed, represented Absolute Reality. Infinity, or ∞, was the myriad manifestations of that Reality. Their mathematical product, ∞ × 0, was not one number, but all numbers, each of which corresponded to individual acts of creation."* The idea perhaps appeared foolish to philosophers, and mathematicians. But Ramanujan found meaning in it.

One friend, Prasanta Chandra Mahalanobis (1893–1972), an eminent statistical scientist and the founder of Indian Statistical Institute in Kolkata, the man who found him shivering in his Cambridge room, later wrote how Ramanujan *"spoke with such enthusiasm about the philosophical questions that sometimes I felt he would have been better pleased to have succeeded in establishing his philosophical theories than in supplying rigorous proofs of his mathematical conjectures."*

To appreciate and realize all that has come out from the life of the genius Ramanujan, one needs to rise much above the elementary state of mind. This will enable one to be more and more amazed, the deeper one probes into what had been going on in his extraordinary mind.

Zero in sciences, engineering, its uses in various countries, and opposition faced

4

4.1 Zero in continuous quantity anywhere is never exact zero

Any continuous quantity such as one liter of honey cannot be measured exactly. When we buy one liter of honey from a supermarket, it is impossible to say that the honey is exactly one liter since, associated with any measuring device, there is always an error. As a matter of fact, it is usually not possible to measure any continuous quantity with an accuracy (complement of the term relative error—observe that absolute error is often not much useful) greater than 0.005%. However, based on the measuring error associated with a measuring device, what we can say is: The exact quantity of honey lies in the interval (1 L − 10 mL, 1 L + 10 mL) or, equivalently (1 L − (1/100) L, 1 L + (1/100) L) based on the accuracy of the measuring device used. This interval should be logically as narrow as possible and also 100% reliable since writing a big interval is evidently useless, although true/reliable.

4.1.1 Physics and engineering

When we say zero in physics or any science/engineering in connection with a continuous quantity, it never implies the absolute (i.e., exact) zero under all physical/practical circumstances; instead, it always implies a numerical zero (never unique). However, if the continuous quantity is completely absent/nonexistent, only then will this quantity be considered exactly zero. The value zero plays a special role for many physical quantities. For some quantities, the zero level is naturally distinguished from all other levels, whereas for others it is more or less arbitrarily chosen.

For example, on the Kelvin temperature scale, zero is the coldest possible temperature (depending on the scale used, negative temperatures exist but are not actually colder), whereas on the Centigrade (Celsius) scale, zero is arbitrarily defined to be at the freezing point of water (the freezing/melting point of water (without any impurity) at 1 atm is 0 °C (32 °F, 273.15 K)). Measuring sound intensity in decibels or phons, the zero level is arbitrarily set at a reference value—for example, at a value for the threshold of hearing. In physics, the zero-point energy is the lowest possible energy that a quantum mechanical physical system may possess and is the energy of the ground state of the system.

Zero: A landmark discovery, the dreadful void, and the ultimate mind. DOI: http://dx.doi.org/10.1016/B978-0-08-100774-7.00004-1

4.1.2 Chemistry

Zero has been proposed as the atomic number of the theoretical element *tetraneutron*. It has been shown that a cluster of four neutrons may be stable enough to be considered an atom in its own right. This would create an element with no protons and no charge on its nucleus.

As early as 1926, Prof. Andreas von Antropoff (1878–1956), a German chemist, coined the term neutronium for a conjectured form of matter made up of only neutrons (with no protons), which he placed as the chemical element of atomic number zero at the head of his new version of the Periodic Table. It was subsequently placed as a noble gas in the middle of several spiral representations of the periodic system for classifying the chemical elements.

4.1.3 Computer science

The most common practice throughout human history has been to start counting at number 1, and this is the practice in early classic Computer (Science) programming languages such as FORTRAN and COBOL. However, in the late 1950s LISP introduced zero-based numbering for arrays while ALGOL 58 introduced completely flexible basing for array subscripts (allowing any positive, negative, or zero integer as the base for array subscripts), and most subsequent programming languages adopted one or the other of these positions.

For example, the elements of an array, say, *M[103]* are numbered starting from 0 in the programming language C, so that for an array of n items the sequence of array indices runs from 0 to n − 1. This permits an array element's location to be calculated by adding the index directly to address of the array, while 1-based indexing languages precalculate the array's base address to be the position one element before the first.

There can be confusion between 0 and 1-based indexing, for example JAVA's JDBC indexes parameters from 1, JAVA itself uses 0-based indexing though.

In databases, it is possible for a field not to have a value. It is then said to have a null value. For numeric fields it is not the value zero. For text fields this is not blank nor the empty string. The presence of null values leads to three-valued logic. No longer is a condition either *true* or *false*, but it can be *undetermined.* Any computation including a null value delivers a null result. Asking for all records with value 0 or value $\neq 0$ will not produce all records, since the records with value null are not included.

A null pointer is a pointer in a computer program that does not point to any object or function. In C, the integer constant 0 is converted into the null pointer during compilation when it appears in a pointer context, and so 0 is a standard way to refer to the null pointer in code. However, the internal representation of the null pointer may be any bit pattern (possibly different values for different data types).

4.1.4 Computational science/mathematics

$-0 = 0 = +0$, that is, −0, 0, and +0 represent exactly the same number, that is, there is no "negative zero" or a "positive zero" distinct from zero. In some signed number representations (but not the two's complement representation used to represent integers in most computers today) and most floating point number representations, zero has two distinct representations, one grouping it with the positive numbers and one with the negative numbers; this latter representation is known as negative zero. From a noncomputational/computer scientist's point of view, who is involved in scientific/engineering computations, one would usually be involved in doing computations using a digital computer where one will mainly (almost always) encounter a *numerical zero* (nonunique, relative) which is distinctly different from the exact zero (unique, absolute).

4.1.5 Algebra

The number 0 is the smallest nonnegative integer. The natural number following the number 0 is 1 and no natural number precedes 0. Zero (0) may/may not be considered a natural number, but it is a whole number, a rational number, a real number, a complex number, and an algebraic number.

Zero is neither a positive nor a negative number. It appears in the middle of a number line (a number line is a picture of a straight line on which every point is assumed to correspond to a real number (infinity of them) and every real number to a point). It is neither a composite number nor a prime number. The number 0 cannot be prime since it has an infinite number of factors. It cannot be composite because it cannot be expressed as the product of prime numbers (0 must always be one of the factors). Zero is, however, an even number.

Rules involving 0 Let x be any real or complex number unless otherwise stated. Then

1. *Addition:* $x + 0 = 0 + x = x$. Here 0 is an identity (neutral) element.
2. *Subtraction:* $x - 0 = x$ and $0 - x = -x$.
3. *Multiplication:* $x \cdot 0 = 0 \cdot x = 0$.
4. *Division:* $0/x = 0$, for nonzero x. But $x/0$ is undefined, because 0 has no multiplicative inverse (no real or complex number multiplied by 0 produces 1). Also, it may be viewed as the violation of the law "Thou shalt not divide by zero."
5. *Exponentiation:* $x^0 = x/x = 1$, except that the case $x = 0$ may be left undefined in some contexts; for all positive real x, $0^x = 0$.
6. *Limit operation:* The expression $^0/_0$, which may be obtained in an attempt to determine the limit of an expression of the form $f(x)/g(x)$ as a result of applying the limit operator independently to both operands of the fraction, is a so-called "indeterminate form." That does not simply mean that the limit sought is necessarily undefined; rather, it means that the limit of $f(x)/g(x)$, if it exists, must be found by another method, such as L' Hôpital's rule (provided "zero" is clearly understood and the rule is appropriately (not mechanically) applied). Consider the evaluation of the expression $(1 - \cos 2x)^{0.5}/x$ when x tends to 0 in the limit. The limit simply does not exist as the left-hand limit is not the same as the right-hand limit.
7. *Factorial operation:* The value of 0! is 1.
8. *Roots of an equation:* A zero of the function f(x) or, equivalently, a root of the equation $f(x) = 0$ is a point x in the domain of the function such that $f(x) = 0$. When there are finitely many zeros these are called the roots of the function. Similarly we have zeros of a holomorphic function. (A holomorphic function is a complex-valued function of one or more complex variables that is complex differentiable in a neighborhood of every point in its domain.)

9. *Zero morphism/map:* The zero function (or zero map) on a domain D is the constant function with 0 as its only possible output value, that is, the function f defined by f(x) = 0 for all x in D. A particular zero function is a zero morphism in category theory. For instance, a zero map is the identity in the additive group of functions. The determinant on noninvertible square matrices is a zero map.

4.1.5.1 Abstract algebra

The symbol 0 is commonly used to denote a zero (neutral) element, for addition (if defined on the structure under consideration) and an absorbing element for multiplication (if defined).

4.1.6 Set, lattice, category, and recursion theories, logic, and beyond

- *Set theory* The number/symbol 0 is the cardinality of the empty set: if one does not have any cows, then one has 0 cows. In fact, in certain axiomatic developments of mathematics from set theory, 0 is *defined* to be the empty set. When this is done, the empty set is the von Neumann cardinal assignment for a set with no elements. The cardinality function, applied to the empty set, returns the empty set as a value, thereby assigning it 0 elements.

  Also 0 is the lowest ordinal number, corresponding to the empty set viewed as a well-ordered set.
- *Lattice theory* The symbol 0 may denote the bottom element of a bounded lattice.
- *Category theory* The symbol 0 is sometimes used to denote an initial object of a category.
- *Recursion theory* The symbol 0 can be used to denote the Turing degree of the partial computable functions.
- *Propositional logic* 0 may be used to denote the truth value false.
- *Some branches of mathematics* Several branches of mathematics have zero elements, which generalize either the property $0 + x = x$, or the property $0 \times x = 0$, or both.

4.1.7 Telephony, DVD, Roulette wheels and Formula One race

- *Telephony* Pressing 0 is often used for dialing outside of an organization network or to a different city or region. Pressing 00 is used for dialing outside a country. In some countries, dialing 0 places a call for operator assistance.
- *DVD* A DVD that can be played in any region is sometimes referred to as being a "region 0" DVD.
- *Roulette wheels* usually feature a "0" or a "00" space whose presence is ignored when calculating payoffs (thus permitting the house to win in the long run).
- *Formula One race* If the reigning World Champion no longer competes in Formula One in the year following his victory in the title race, 0 is given to one of the drivers of the team that the reigning champion won the title with. This happened in 1993 and 1994, with *Damon Graham Devereux Hill* (born 1960, a retired British racing driver from England) driving car 0, due to the reigning World Champions *Nigel Ernest James Mansell* (born 1953) and *Alain Marie Pascal Prost* (born 1955), respectively, not competing in the championship.

4.1.8 Tolerance

A zero tolerance policy imposes automatic punishment for infractions of a stated rule, with the intention of eliminating undesirable conduct. Zero-tolerance policies forbid persons in positions of authority from exercising discretion or changing punishments to fit the circumstances subjectively; they are required to impose a pre-determined punishment regardless of individual culpability, extenuating circumstances, or history. This pre-determined punishment need not be severe, but it is always meted out.

An *engineering tolerance* is the permissible limit or limits of variation in a physical dimension or a measured value of a material or measured values of parameters such as temperature and humidity or the space between a bolt and a nut (in mechanical engineering). *Tolerance* in engineering is always greater than zero, ideally it should be zero though.

4.2 Uses of zero in various countries with stress on Indian zero and its transmission

- *Babylon* was an Akkadian city-state (founded in 1894 BC by an Amorite dynasty) of ancient Mesopotamia, the remains of which are found in present-day Hillah, Babylon Province, Iraq, about 85 km (53 miles) south of Baghdad.

If zero was merely a place holder symbol, indicating the absence of a magnitude at a specified place/position (e.g., the zero in 101 indicated the absence of any "tens" in one hundred and one), then such a zero was already present in the Babylonian number system. Babylonians in Mesopotamia (3000 BC) had *a sexagesimal system* using base 60. Greeks and Romans had a cumbersome system. Numbers are formed by combining symbols together and adding the values. For example, MMXIII is $1000 + 1000 + 10 + 1 + 1 + 1 = 2013$. Generally, symbols are placed in order of value, starting with the largest values. When smaller values precede larger values, the smaller values are subtracted from the larger values, and the result is added to the total. For instance, MCMXLVX $= 1000 + (1000 - 100) + (50 - 10) + (10 - 5) = 1945$. Roman numerals, as used today, are based on seven symbols viz. I = 1, V = 5, X = 10, L = 50, C = 100, D = 500, and M = 1000.

If zero was represented by just an empty space within a well-defined positional number system, then such zero seemed to be present in *Chinese mathematics* during the first millennium AD.

Several civilizations had used zero to signify nothing. For instance, if you have four cows and they all die, then you are left with nothing. However, *the Indians* were the first to see that *zero can be used to specify not only nothing in a context but also something beyond it.* At different places in a number, it adds different values. For example, 83 is different from 803, 8003, 830 and so on.

The zero viewed and used by the Indians was a nonambiguously defined multipurpose mathematical object: (i) a number, (ii) a symbol, (iii) a direction separator, (iv) a magnitude, and (v) a place holder, five-in-one operating with a fully established positional number system. The Indian zero is now the universal zero and is time-invariant, and unique. This is in contrast to local or, equivalently, relative or, equivalently, numerical zero.

One might think that when a place-value number system came into existence, the idea of 0 as an empty place indicator was necessary, yet the Babylonians had a place-value number system without this feature for over one millennium. Further there is no evidence that the Babylonians felt that there was any problem with the ambiguity which existed. Notably, original texts survive from the era of Babylonian mathematics. The Babylonians wrote on tablets of unbaked clay, using cuneiform writing. The symbols were pressed into soft clay tablets with the slanted edge of a stylus and so had a wedge-shaped appearance (and hence the name cuneiform). Many tablets from around 1700 BC did not perish. We could scan the original texts.

However, their notation for numbers was different from that of ours (based on 60 and not on 10) but to translate into our notation they would not distinguish between 4207 and 427. Only the context would have to demonstrate which was intended. It was not until around 400 BC that the Babylonians put two wedge symbols into the place where we would put zero to indicate which was meant, 427 or 42″7.

The two wedges were not the only notation used. However, on a tablet found in Kish, an ancient Mesopotamian city located east of Babylon in what is today south-central Iraq, a different notation was employed. This tablet, thought to date from around 700 BC, used three hooks to denote an empty place in the positional notation. Other tablets dated from around 700 BC used a single hook for an empty place. There is one common feature to this use of different marks to denote an empty position. This is the fact that it never occurred at the end of the digits but always between two digits. So although we find 42″7 we never find 427″. One has to assume that the older feeling that the context was sufficient to indicate which was intended still applied in these cases.

If this reference to context does not sound appealing or seems not logical, then it is worth noting that we still use context to interpret numbers today. If I take a bus to a nearby city and ask what the fare is then I know that the answer *"It's four twenty" means four dollars twenty cents.* Yet if the same answer is given to the question about the cost of a flight from Los Angeles to Orlando, then I know that *four hundred and twenty dollars* is what is intended.

We can see from this that the early use of zero to denote an empty place is not really the use of zero as a number at all, merely the use of *some type of punctuation mark* so that the numbers had the correct interpretation.

- *Greece* The ancient Greeks started contributing to mathematics around the time when zero as an empty place indicator was coming into use in Babylonian mathematics. The Greeks did not adopt a positional number system. It is worth pondering just how significant this fact was. How could the brilliant mathematical advances of the Greeks not see them adopt a number system with all the advantages that the Babylonian place-value system possessed? The real answer to this question is more subtle than the simple answer given below.

The Greek mathematical achievements were *geometry-oriented. Euclid's Elements* consisting of 13 books is a geometry (study of shape) based mathematical treatise written by the Greek mathematician Euclid in Alexandria in 300 BC. Euclid, often referred to as the *Father of Geometry*, was active in Alexandria during the reign of Ptolemy I Soter I also known as Ptolemy Lagides (367–283 BC). (Ptolemy I Soter I was a Macedonian general under Alexander the Great (Alexander III of Macedon (356–323 BC)), who became ruler of Egypt during 323–283 BC and founder of both the Ptolemaic Kingdom and the Ptolemaic Dynasty. In 305–304 BC he took the title of Pharaoh.) One of the 13 books, is on number theory. In essence, Greek mathematicians were not required to name their numbers since they worked with numbers as lengths of lines. However, numbers which needed to be named for records were used by merchants, not mathematicians, and thus no clever notation was called for.

The mathematical and astronomical tablets from the Seleucid era (i.e., after the time of Alexander the Great) employed a numerical system which included a symbol for zero. The zero occurs either at the beginning of a number or within it, but never at the end.

In the third century BC, during the reign of the Buddhist Emperor Asoka (Asoka or Ashoka Maurya (304–232 BCE), also known as Ashoka the Great, was an Indian emperor of the Maurya Dynasty), two different systems of numerals were in use among Hindu mathematicians. One was Kharosti (Kharosti or Kharosthi script included a set of numerals that are reminiscent of Roman numerals), while the other was Brahmi. It was in the sixth or seventh century AD that the system of numerals that we use today, positional decimal arithmetic including a zero, first appeared in India.

Now there were exceptions to what just has been told. The exceptions were the mathematicians who were involved in recording astronomical data. Here we find the first use of the symbol which we recognize today as the notation for zero, for Greek astronomers began to use the symbol O. There are many theories why this particular notation was used. Some historians favor the explanation that it is omicron, the first letter of the Greek word for nothing namely "ouden." Neugebauer, however, dismisses this explanation since the Greeks already used omicron as a number (ø)—it represented 70 (the Greek number system was based on their alphabet). Other explanations offered include the fact that it stands for "*obol*," *a coin of almost no value*, and that it arises when counters were used for counting on a sand board. The suggestion here is that when a counter was removed to leave an empty column it left a depression in the sand which looked like O.

Claudius Ptolemy (90–168 AD), a mathematician, astronomer, geographer, astrologer, poet of a single epigram in the Greek Anthology, and Greco-Roman citizen of Egypt, who wrote in Greek in the *Almagest*, a mathematical and astronomical treatise on the apparent motions of the stars and planetary paths written around 130 AD, uses the Babylonian *sexagesimal system* together with the empty place holder O. By this time Claudius is using the symbol both between digits and at the end of a number and one might tend to believe that at least zero as an empty place holder had been firmly established. This, however, was not what had happened. *Only a few astronomers used the notation and it had fallen out of use a number of times before finally establishing itself permanently.* The idea of the zero place (most probably not thought of as a number by Claudius who seemed to still consider it as a sort of punctuation mark) makes its next appearance in Indian mathematics.

- *India* The scene now moves to India where it is fair to say the numerals and number system were born. These have evolved into the highly sophisticated ones that we use today. Of course that is not to say that the Indian system did not owe something to earlier systems and many historians of mathematics believe that the Indian use of zero evolved from its use by Greek astronomers. There are some historians who seem to want to play down the contribution of the Indians in a most unreasonable way, there are also those who make claims about the Indian invention of zero which seem to go far too far. For instance Mukherjee claims

> *... the mathematical conception of zero ... was also present in the spiritual form from 17 000 years back in India.*

What is certain is that by around 650 AD the use of zero as a number came into Indian mathematics. The Indians also used a place-value system and zero was used to denote an empty place. In fact there is evidence of an empty place holder in positional numbers from as early as 200 AD in India but some historians dismiss these as later forgeries. Let us examine this latter use first since it continues the development described above.

The concept of zero as a number and not merely a symbol for separation is attributed to India, where, the ninth century AD practical calculations were carried out using zero, which was treated like any other number, even in case of division. The Indian scholar *Pingala* (about 500 BC) was the author of the *Chhandah-shastra*, the Sanskrit book on meters, or long syllables. This Indian mathematician was from the region that is present day Kerala state in India (circa fifth–second century BC). He used *binary numbers* in the form of short and long syllables (the latter equal in length to two short syllables), making it similar to *Morse code*. He and his contemporary Indian scholars used the Sanskrit word *śūnya* to refer to zero or *void*. The use of a blank on a counting board to represent 0 dated back in India to the fourth century BC. *It may be noted from Chapters 1 and 3 that the modern decimal-oriented place-value system originated in the third millennium BC due to Aryabhatta.*

The text to use a decimal place-value notation (more than three millennia after Aryabhatta) is the Jain text entitled the *Lokavibhaga*, dated 458 AD, where *shunya* ("void" or "empty") was employed for this purpose as mentioned earlier in this section.

Zero, represented Absolute Reality. Infinity, or ∞, was the myriad manifestations of that Reality. Their mathematical product, $\infty \times 0$, was not one number, but all numbers, each of which correspond to individual acts of creation. Zero and negative numbers, for instance, are impossible in Greek geometry, for a line cannot be zero or less in length, for then it is no longer a line. The Greeks never even invented a symbol for zero. They had no use for it. Their number field included only real, positive integers, and when they did discover another kind of number, they retreated in horror.

The word "shunya" to express "void" or "absence," that the Sanskrit language already possessed is synonymous with "vacuity," this word had for several centuries constituted the central element of a mystical and religious philosophy which had become a way of thinking.

Thus words such as abhra, ambara, antariksha, gagana, kha, nabha, vyant, or vyoman, which literally meant the sky, space, the atmosphere, the firmament, or the canopy of heaven, came to signify not only a void, but also zero. There was also the word akasha, the principal meaning of which was "ether," the last and the most subtle of the "five elements" of Hindu philosophy, the essence of all that is believed to be uncreated and eternal, the element which penetrates everything, the immensity of space, even space itself.

Sanskrit and corresponding English words concerning zero and infinity and significance of void and zero are as follows. Literally "void" is the principal Sanskrit term for "zero." However, the Sanskrit language (the excellent literary instrument of mathematicians, astronomers, and all Indian scholars) has many synonyms for expressing this concept. The list includes the words Abhra (atmosphere), absence (Shunya), Akasha (space, ether or element which penetrates everything), Ambara (atmosphere), Ananta (immensity of space, infinity, the serpent of eternity), Antariksha (atmosphere, space), Bindu (point or dot), Gagana (canopy of heaven, the firmament), Jaladharapatha (Voyage on water), Kha (space, sky), Nabha (sky, atmosphere), Nabhas (sky, atmosphere), Purna (entire, complete, the full, the fullness, totality, integrity, completion, the state of that which is entire, finished, complete), Randhra (hole), Shunya (void, absence, nothingness, nothing, the insignificant, the negligible quantity, nullity), Vindu (point or dot), Vishnupada (foot of Vishnu, the zenith), Vyant (sky, space), Vyoman (sky, space).

To the Indian mind, space was the "void" which had no contact with material objects, and was an unchanging and eternal element which defied description; thus the association between the elusive character and very different nature of zero (as regards numerals and ordinary numbers) and the concept of space was immediately obvious to the Indian scholars.

The association between ether and "void" is also obvious because akasha (to the Indian mind) is devoid of all substance, being considered the condition of all corporal extension and the receptacle of all substances formed by one of the other four elements (earth, water, fire, and air). In other words, once zero had been invented and put into use, it brought about the realization that, in terms of existence, akasha played a role comparable with the one which zero performed in place-value system, in calculations, in mathematics, and in the sciences.

In drawings and pictograms, the canopy of heaven is universally represented either by a semi-circle or by a circular diagram or by a whole circle. The circle has always been regarded as the representation of the sky and of the Milky Way as it symbolizes both activity and cyclic movements. Thus the little circle, through a simple transposition and association of ideas, came to symbolize the concept of zero for the Indians.

The point is the most insignificant geometrical figure, constituting as it does the circle reduced to its simplest expression, its center. The bindu represents the universe in its nonmanifest form, the universe before it was transformed into the world of appearances (rupadhatu). According to Indian philosophy, this uncreated universe possessed a creative energy, capable of generating everything and anything: it was the causal point. The most elementary of all geometric figures, which is capable of creating all possible lines and shapes (rupa) was thus associated with zero, which is not only the most negligible of quantities, but also, and above all, the most fundamental of all abstract mathematics. The point was thus used to represent zero, most notably in the Sharada system of Kashmir, and in the vernacular notations of Southeast Asia.

From the fifth century CE, the Indian zero, in its various forms, already surpassed the heterogeneous notions of vacuity, nihilism, nullity, insignificance, absence, and nonbeing of Greek–Latin philosophies. Shunya embraced all these concepts, following a perfect homogeneity: it signified not only void, space, atmosphere, and ether, but also the noncreated, the nonproduced, nonbeing, nonexistence, the unformed, the unthought, the nonpresent, the absent, nothingness, nonsubstantiality, nothing much, insignificance, the negligible, the insignificant, nothing, nil, nullity, unproductiveness, of little value, and worthlessness.

It was also, and above all, an eminently abstract concept: in the simplified Sanskrit system, as well as in the positional system of the numerical symbols, the word shunya and its various synonyms served to mark the absence of units within a given decimal order, in a medial position as well as in an initial or final position; the point or the little circle were used in the same way.

This zero was also a mathematical operator: if it was added to the end of a numerical representation, the value of the representation was multiplied by 10. By freeing the nine basic numerals from the abacus and inventing a sign for zero, the Indian scholars made significant progress, primarily simplifying quite considerably the rules of a technique which would lead to the birth of our modern written calculation.

The Indian people were the only civilization to take the decisive step toward the perfection of numerical notation. We owe the discovery of modern numeration and the elaboration of the very foundations of written calculations to India alone. It is likely that this important historical event took place around the fourth century CE. Thanks are due to the genius of the Indian arithmeticians that three significant ideas were combined:

1. nine numerals which gave no visual clue as to the numbers they represented and which constituted the prefiguration of our modern numerals,
2. the discovery of the place-value system, which was applied to these nine numerals, making them dynamic numerical signs, and
3. the invention of zero and its enormous operational potential.

Thus we can see that the Indian contribution was essential because it united calculations and numerical notation, enabling the *democratization* of calculation. For thousands of years this field had only been accessible to the privileged few (professional mathematicians). These discoveries made the domain of arithmetic accessible to anyone. It still remained for the Indian scholars to perfect the concept of zero and enrich its numerical significance.

Beforehand, the shunya had only served to mark the absence of units in a given order. The Indian scholars, however, soon filled in the gap. Thus, in a short space of time, the concept became synonymous with what we now refer to as the "number zero" or the "zero quantity." The shunya was placed amongst the Samkhya, which means it was given the status of a "number."

Astronomer Varahamihira (working 123 BC), in *Pancha Siddhantika*, mentioned the use of zero in mathematical operations, as did Bhaskhara (before 123 BC) in his commentary on the Aryabhatiya. In *Brahmasphuta Siddhanta*, Brahmagupta (born 30 BC) defined zero as the result of the subtraction of a number by itself ($a - a = 0$), and described its properties in the following terms:

When zero (shunya) is added to a number or subtracted from a number, the number remains unchanged; and a number multiplied by zero becomes zero. Moreover, in the same text, Brahmagupta gives the following rules concerning operations carried out on what he calls "fortunes" (dhana), "debts" (rina), and "nothing" (kha):

A debt minus zero is a debt.
A fortune minus zero is a fortune.
Zero (shunya) minus zero is nothing (kha).
A debt subtracted from zero is a fortune.
So a fortune subtracted from zero is a debt.
The product of zero multiplied by a debt or a fortune is zero.
The product of zero multiplied by itself is nothing.
The product or the quotient of two fortunes is one fortune.
The product or the quotient of two debts is one fortune.
The product or the quotient of a debt multiplied by a fortune is a debt.
The product or the quotient of a fortune multiplied by a debt is a debt.

The foregoing "fortunes–debt–nothing" statements are essentially the same as his rules regarding addition, subtraction, and multiplication stated earlier.

Modern algebra was born, and the mathematicians had thus formulated the basic rules: by replacing "fortune" and "debt" respectively with "positive number" and "negative number," we can see that at that time the Indian mathematicians knew the famous "rule of signs" as well as all the fundamental rules of algebra.

It is clear how much we owe to this brilliant civilization, and not just in the field of arithmetic; by opening the way to the generalization of the concept of the number, the Indian scholars enabled the rapid development of algebra, and thus played an essential part in the development of mathematics and exact sciences. The discoveries of these men doubtless required much time and imagination, and above all a great ability for abstract thinking. The reader will not be surprised to learn that these major discoveries took place within an environment which was at once mystical, philosophical, religious, cosmological, mythological, and metaphysical.

One should not be surprised to learn that, a thousand years earlier than the Europeans, Indian mathematicians already knew that zero and infinity were inverse concepts. They realized that when any number is divided by zero the result is infinity: $a/0 = \infty$, this "quantity" (in magnitude) undergoing no change if it is added to or subtracted from a finite number.

Null is opposite of the unlimited. For thousands of years, people stumbled along with inadequate and useless systems which lacked a symbol for "empty" or "nothing." Similarly there was no way of conceiving of "negative" numbers $-1, -2, -3, \ldots$, such as we nowadays use routinely to express, for example, sub-zero temperatures or bank accounts in deficit. Therefore a subtraction such as 3–5 was for a long time considered to be impossible. We have seen how the discovery of zero swept away this obstacle so that ordinary (natural) numbers were extended to include their "mirror images" with respect to zero. That inspired and difficult invention, zero, gave rise to modern algebra and all the branches of mathematics which have come about since the Renaissance.

We now come to consider the first appearance of zero as a number. Let us first note that it is not in any sense a natural candidate for a number. Since early times numbers are words which refer to collections of objects. Certainly the idea of number became more and more abstract and this abstraction then made possible the consideration of zero and negative numbers which do not crop up as properties of collections of objects. Of course the problem which arises when one attempts to consider zero and negatives as numbers is how they interact in regard to the arithmetic operations, that is, addition, subtraction, multiplication, and division. In three important books the Indian mathematicians Brahmagupta, Mahavira (817–878 AD), and Bhaskara attempted to address this problem.

As stated earlier in this section, in 830 AD, around 800 years after Brahmagupta's famous book *Brahmasphuta Siddhanta*, *Mahavira* wrote *Ganita Sara Samgraha* dated 850 AD, as an *update* of Brahmagupta's book, in which Mahavira correctly states … a number multiplied by zero is zero, and a number remains the same when zero is subtracted from it.

Filliozat writes

> *This book (Ganita Sara Samgraha) deals with the teaching of Brahmagupta but contains both simplifications and additional information. … Although like all Indian versified texts, it is extremely condensed, this work, from a pedagogical point of view, has a significant advantage over earlier texts.*

It consisted of nine chapters and included all mathematical knowledge of mid-ninth century India. It provides us with the bulk of knowledge which we have of Jaina mathematics and it can be seen as in some sense providing an account of the work of those who developed this mathematics. There were many Indian mathematicians before the time of Mahavira but, perhaps surprisingly, their work on mathematics is always contained in texts which discuss other topics such as astronomy. The *Ganita Sara Samgraha* by Mahavira is an earliest Indian text that we possess and which is devoted entirely to mathematics.

In the introduction to the work Mahavira paid tribute to the mathematicians whose work formed the basis of his book. These mathematicians included Aryabhata, Bhaskara I, and Brahmagupta. Mahavira writes:

> *With the help of the accomplished holy sages, who are worthy to be worshipped by the lords of the world … I glean from the great ocean of the knowledge of numbers a little of its essence, in the manner in which gems are picked from the sea, gold from the stony rock and the pearl from the oyster shell; and I give out according to the power of my intelligence, the Sara Samgraha, a small work on arithmetic, which is however not small in importance.*

The nine chapters of the *Ganita Sara Samgraha* are

1. Terminology
2. Arithmetical operations
3. Operations involving fractions
4. Miscellaneous operations
5. Operations involving the rule of three
6. Mixed operations

7. Operations relating to the calculations of areas
8. Operations relating to excavations
9. Operations relating to shadows

Throughout the work a place-value system with nine numerals is used or sometimes Sanskrit numeral symbols are used. Of interest in Chapter 1 regarding the development of a place-value number system is Mahavira's description of the number 12345654321 (a palindrome) which he obtains after a calculation. He describes the number as … *beginning with one which then grows until it reaches six, then decreases in reverse order.*

Observe that this wording makes sense to us using a place-value system but would not make sense in other systems. It clearly demonstrates that Mahavira is comfortable with the place-value number system.

Among topics Mahavira discussed in his treatise was arithmetic operations with fractions including methods to decompose integers and fractions into unit fractions. For instance,

$$3/13 = 1/6 + 1/26 + 1/39.$$

He examined methods of squaring numbers which, although a special case of multiplying two numbers, can be computed using special methods. He also discussed integer solutions of first degree indeterminate equation by a method called *kuttaka*. The kuttaka (or the "pulveriser") method is based on the use of the Euclidean algorithm but the method of solution also resembles the continued fraction process of Euler given in 1764. The work kuttaka, which occurs in many of the treatises of Indian mathematicians of the classical period, has taken on the more general meaning of "algebra."

An example of a problem given in the *Ganita Sara Samgraha* which leads to indeterminate linear equations is the following.

> *Three merchants find a purse lying in the road. One merchant says "If I keep the purse, I shall have twice as much money as the two of you together." "Give me the purse and I shall have three times as much" said the second merchant. The third merchant said "I shall be much better off than either of you if I keep the purse, I shall have five times as much as the two of you together." How much money is in the purse? How much money does each merchant have?*

If the first merchant has x, the second y, the third z, and p is the amount in the purse then, according to the condition of the problem,

$$p + x = 2(y + z), p + y = 3(x + z), p + z = 5(x + y).$$

which is a system of three consistent linear (algebraic) equations in four unknowns x, y, z, and p having infinity of solutions. The smallest positive integral solution is $p = 15, x = 1, y = 3, z = 5$. Any solution in positive integers is a multiple of this solution as Mahavira rightly claimed. For instance, $p = 30, x = 2, y = 6, z = 10$ is also a positive integral solution.

Mahavira gave special rules for the use of permutations and combinations which was a topic of special interest in Jaina mathematics. He also described a process for calculating the volume of a sphere and one for calculating the cube root of a number. He looked at some geometrical results including right-angled triangles with rational sides.

Mahavira also attempted to solve certain mathematical problems which had not been studied by other Indian mathematicians. For example, he gave an approximate formula for the area and the perimeter of an ellipse. In all the foregoing arithmetic and algebra, zero did not pose any problem until a division by zero is encountered. Obviously this is compatible in today's arithmetic and algebra. We know that division by zero is completely forbidden under any circumstances at all time in the past, present, and future. Hayashi writes

> *The formulas for a conch-like figure have so far been found only in the works of Mahavira and Narayana.*

It is reasonable to ask what a "conch-like figure" is. It is two unequal semicircles (with diameters AB and BC) stuck together along their diameters. Although it might be reasonable to suppose that the perimeter might be obtained by considering the semicircles, Hayashi claims that the formulas obtained ... *were most probably obtained not from the two semicircles AB and BC.*

However his attempts to improve on Brahmagupta's statements on dividing by zero seem to lead him into error. He (Mahavira) writes, "A number remains unchanged when divided by zero." Since this is clearly incorrect, my use of the words "seem to lead him into error" might be seen as confusing. The reason for this phrase is that some commentators on Mahavira have tried to find excuses for his incorrect statement.

However, in all the foregoing arithmetic and algebraic rules stated by Mahavira, zero did not pose any problem anywhere except when a division by zero is met with. This is evidently noncontradictory in arithmetic and algebra that we practise today. We know that division by zero is completely forbidden under any circumstances at all time in the past, the present, and the future.

- *Central America* Perhaps we should note at this point that there was another civilization which developed a place-value number system with a zero. This was the Maya people who lived in central America, occupying the area which today is southern Mexico, Guatemala, and northern Belize. This was an old civilization but flourished particularly between 250 AD and 900 AD. We know that by 665 AD they used a place-value number system to base 20 with a symbol for zero. However their use of zero goes back further than this and was in use before they introduced the place-valued number system. This is a remarkable achievement but sadly did not influence other peoples.

4.2.1 Indian mathematics reached west through Arabs and east to China directly

The brilliant work of the Indian mathematicians was transmitted to the Islamic and Arabic mathematicians further west. It came at an early stage for al-Khwarizmi wrote

Al'Khwarizmi on the Hindu Art of Reckoning which describes the Indian place-value system of numerals based on 1, 2, 3, 4, 5, 6, 7, 8, 9, and 0. This work was the first in what is now Iraq to use zero as a place holder in positional base notation. *Ibn Ezra*, in the twelfth century, wrote three treatises on numbers which helped to bring the Indian symbols and ideas of decimal fractions to the attention of some of the learned people in Europe.

The Book of the Number describes the decimal system for integers with place values from left to right. In this work ibn Ezra uses zero which he calls galgal (meaning wheel or circle). Slightly later in the twelfth century AD *al-Samawal* (Ibn Yahyā al-Maghribī al-Samawal was a Muslim mathematician, astronomer, and physician. Though born to a Jewish family, he concealed his conversion to Islam for many years in fear of offending his father, then openly embraced Islam in 1163 after he had a dream telling him to do so. His father was a Rabbi from Morocco.) writes

> *If we subtract a positive number from zero the same negative number remains. ... if we subtract a negative number from zero the same positive number remains.*

The Indian ideas spread east to China as well as west to the Islamic countries. In 1247 AD the Chinese mathematician *Ch'in Chiu-Shao* wrote *Mathematical treatise in nine sections* which uses the symbol O for zero. A little later, in 1303AD, *Zhu Shijie* wrote *Jade mirror of the four elements* which again uses the symbol O for zero.

Fibonacci was one of the main people to bring these new ideas about the number system to Europe. *Pogliani* et al. writes

> *An important link between the Hindu–Arabic number system and the European mathematics is the Italian mathematician Fibonacci.*

In *Liber Abaci* he described the nine Indian symbols together with the sign 0 for Europeans in around 1200 AD but it was not widely used for a long time after that. It is significant that Fibonacci is not bold enough to treat 0 in the same way as the other numbers 1, 2, 3, 4, 5, 6, 7, 8, 9 since he speaks of the "sign" zero while the other symbols he speaks of as numbers. Although clearly bringing the Indian numerals to Europe was of major importance we can see that in his treatment of zero he did not reach the sophistication of the Indians Brahmagupta, Mahavira, and Bhaskara nor of the Arabic and Islamic mathematicians such as *al-Samawal*.

4.2.2 Zero still had to pass through significant opposition

One might have thought that the progress of the number systems in general, and zero in particular, would have been steady from this time on. However, this was far from the case. *Cardan* (Girolamo Cardan, 1501–1576) solved cubic and quartic equations *without using zero*. He would have found his work in the 1500s so much easier if he had had a zero but it was not part of his mathematics. By the 1600s zero began to come into widespread use but still only after encountering a lot of resistance.

Of course there are still signs of the problems caused by zero. Recently many people throughout the world celebrated the new millennium on January 1, 2000 AD. Of course they celebrated the passing of only 1999 years, since when the calendar was set up no year zero was specified. Although one might forgive the original error, it is a little surprising that most people seemed unable to understand why the third millennium and the twenty-first century begin on January 1, 2001 AD. *Zero is still causing problems!*

4.3 Y2K problem

Before falling from grace by admitting to India's biggest accounting fraud, B Ramalinga Raju, founder of Satyam Computer Services Limited (currently Mahindra Satyam Indian IT Services Company), was a poster boy of the Indian IT industry who rose to fame with a solution for the highly-feared Y2K crisis at the turn of twentieth century for the entire world. Interestingly, his first business venture was a spinning and weaving mill, named Sri Satyam, while his undoing came in form of spinning a fraud to the tune of Rs 7000 crore, that is, US $ 1166666667 or approximately US $ 1.17 billion (taking Rs. 60 = US $ 1 as in April, 2015) at Satyam Computer—a company he founded and nurtured, although with inflated profits and falsified books. In April 2015, 60-year-old Raju, who was sentenced by a CBI court on April 9/10, 2015 to 7 years in jail with a fine of Rs 5.5 crore, that is, about US $ 0.92 million, founded Satyam in 1987 and made it into what came to be known as India's fourth largest IT firm, which reaped huge profits after making *software solutions to tackle the famous Y2K crisis*. It was feared at that time that a worldwide bug would crash computer systems across the globe, due to abbreviating a 4-digit year to two digits, on January 1, 2000, as the systems did not recognize the year 2000 from the year 1900.

Conclusions

5

5.1 Place-value system beyond 1000 AD: used extensively in India and in use today all over

In our decimal number system, the value of a digit depends on its place, or position, in the number. The place-value system has existed since the Vedic era (7000–4000 BC). Aryabhatta (born 2765 BC) had used it around 2700 BC. There are umpteen references to the usage of this system after 1000 AD and it continues to be used today and will continue to remain in use possibly indefinitely for years/centuries to come (until a better system is evolved).

5.1.1 Shripati

Shripati was an Indian astronomer, who flourished during circa 1039 CE. His works notably include a text entitled *Siddhantashekhara*, in which the place-value system of the Sanskrit numerical system is used frequently. Also the works of *Nilakanthan Somayaji* (around 1444–1544 AD), a distinguished mathematician, include *Siddhantadarpana*, in which the place-value system with Sanskrit numerical symbols is used frequently. Furthermore, the works of *Kamalakara*, another Indian astronomer of the seventeenth century CE, notably include *Siddhantatattvaviveka,* in which the place-value system with Sanskrit numerical symbols is frequently used. The place-value system continues to be used and perhaps will continue to be used indefinitely for years to come since so far no better system has evolved in spite of the best efforts of mathematical and computational scientists.

5.2 Irrational number without zeros among its digits is inconceivable

An irrational number (nonrecurring, i.e., no pattern in its decimal form; in other words, when the decimal form has no pattern whatsoever, it is irrational. If there is a pattern, then it is a good indication for rational) without zeros among its digits is inconceivable. An irrational number (a number that cannot be expressed as the ratio of two integers) will always have zeros in its decimal (or any other radix) representation. It is a conjecture to us now, but the proof should not possibly be difficult.

Any of the irrational numbers such as pi (the ratio of the circumference and the diameter of any circle), e (exponential function of argument 1), and phi (golden ratio) will have zeros in their representation in any number system of any positive integral (finite) radix which has zero (practically always the case) in its digits. It is observed that in each of pi, e, and phi (each having infinity of digits), the 10 digits 0, 1, 2, …, 9

Zero: A landmark discovery, the dreadful void, and the ultimate mind. DOI: http://dx.doi.org/10.1016/B978-0-08-100774-7.00005-3

are uniformly randomly distributed. Consider, for instance, the first 10^5 (i.e., 100,000) digits of pi. Out of these digits, each digits 0, 1, 2, 3, 4, 5, 6, 7, 8, and 9 will occur 10^4 (i.e., 10,000) ±5% (say) times.

Of course, digits of an irrational number may be more (or less) uniformly randomly distributed than another. This uniform distribution strongly suggests to us the *conjecture* "Any irrational (decimal) number without zeros among its digits is nonexistent."

5.3 Two ways of looking at absolute zero

There are basically two ways of looking at zero: (i) as a conventional entity in a/ any number system (specifically, decimal and binary) which everybody assimilates and uses and (ii) as a void, specifically universal void, that is, a state/situation in which nothing is existing including one's own body, which a few people, that is, a few intense thinkers, attempt to experience with the realization of utter dread or with that of ultimate truth/self/God/Consciousness. The second way of experiencing zero occurs in an extraordinary spiritual plane and is exclusive to a few seekers of supreme absolute truth (usually outside the domain of majority of common thinkers/humans). In the domain of numerical computation/computer (finite precision machine), the zero used is usually a numerical zero most of the time, however, sometimes it is the absolute zero.

5.4 Concept of zero existed before Christian era

Zero was regarded as a number (and not merely a symbol or an empty space for separation) in India … whereas the Chinese employed a vacant position. No Long Count date actually using the number 0 has been found before the third century AD, but since the Long Count system would make no sense without some place-holder, and since Mesoamerican glyphs do not typically leave empty spaces, these earlier dates are taken as indirect evidence that the concept of 0 already existed at the time although in the BC/AD scheme there is no year zero. After December 31, 1 BC came January 1, 1 AD. … One may use the astronomer's numbering scheme with negative year numbers if one is not comfortable with the no-year-zero scheme. As discussed in previous chapters, nobody has been able to convincingly claim the date from when the concept of zero came into existence.

5.5 Existence of year zero in astronomical counting is advantageous and preserves compatibility with significance of AD, BC, CE, and BCE

Year zero does not exist in the Anno Domini system usually used to number years in the Gregorian calendar and in its predecessor, the Julian calendar. In this system, the year 1 BC is followed by 1 AD. However, there is a year zero in astronomical year numbering (where it coincides with the Julian year 1 BC) and in ISO 8601:2004 (where it coincides with the Gregorian year 1 BC) as well as in all Buddhist and Hindu calendars. *Astronomical year numbering* is based on AD/CE year numbering, but follows normal decimal integer numbering more strictly. Thus, it has a year 0, the years before that are designated with negative numbers and the years after that are designated with positive numbers.

The suffixes AD, CE (Common Era), BC (Before Christ), or BCE (Before Common Era) are dropped. The year 1 BC/BCE is numbered 0, the year 2 BC is numbered −1, and in general the year n BC/BCE is numbered 1−n. The numbers of AD/CE years are not changed and are written with either no sign or a positive sign; thus in general n AD/CE is simply n or +n. For normal calculation a number zero is often needed, here notably when computing the number of years in a period that spans the epoch (epoch serves as a reference point from which time is measured); the end years need only be subtracted from each other. Inclusion of year zero is thus advantageous and compatible in contexts other than determining a period of time readily in astronomical numbering.

5.6 Zero as a place-holder in Long Count dates

The Olmec civilization (an ancient civilization in south-central Mexico (modern-day states of Veracruz and Tabasco)) flourished during the formative period of Mesoamerica (around 1500–400 BCE). *The Olmecs had the idea of the number zero as a concept as well as a place-holder. This can be seen from the Long Count calendar that required the use of zero as a place-holder within its modified radix-20 (radix-20 cum radix 18) positional number system.* They used a shell glyph as a zero symbol for these Long Count dates, the second oldest of which, on the back of Olmec Stela C at Tres Zapotes, has a date *7.16.6.16.18* that translates to *September, 32 BCE* (Julian).

5.7 Representation of any information: minimum two symbols are required—zero is one of them

Any information—scientific or not—needs at least two symbols for its representation. A blank, if used, is logically considered to be a symbol. In a digital computer, the most stable states are two. On a magnetic medium, a spot is magnetized representing, say 1 or *not* magnetized representing zero which is the best way of representing *not* or, equivalently, *nothing*. A blank (when considered as a symbol) taking the place of a zero will not be uniformly applicable everywhere (unlike a zero) because of ambiguity and possibly context-dependency posed to a human being. More than two stable physical states with fast switching activity representing numbers in base 3 or more are yet to be found out. Every digital computer is only binary in its basic hardware form.

In 1958, under the leadership of Sergei Lvovich Sobolev (1908–1989) and Nikolay Brusentsov (born 1925), the world's first balanced ternary (using three–valued ternary logic instead of two–valued binary logic) computer, *Setun*, was developed and built at Moscow State University. Base 3 or more computers are not as stable as base 2 (binary) computers because of the required stability and fast errorless switching problem (existing in silicon technology in the physical world). The *Setun* computer was built to fulfill the needs of Moscow State University and was manufactured at the Kazan Mathematical plant. Fifty computers were built until production was halted in 1965.

In the period between 1965 and 1970, a regular binary computer replaced it at Moscow State University. Although this replacement binary computer performed equally well, it cost 2.5 times as much as the Setun. In 1970, although a new ternary computer, the Setun–70, was designed, it was soon realized that ternary or higher radix computer, although requiring less writing/printing space and consequently less hardcopy printing (possibly less ink and less wear and tear) costs, cannot be a substitute of a binary computer. For the last five decades, only the binary digital computers have been ruling the world.

Whether a computer is binary or ternary or higher radix, zero is always considered/assumed one of the two or more symbols used to represent a number not only for our convenience but also for our enormous intimacy with and understanding of zero. Any other symbol (not physically implying zero and all its characters) will serve no better purpose in our minds, which are highly conditioned with the symbol of a zero (different languages usually use different symbols for zero although a round "Oh" or an oval-shaped "Oh" is the most commonly used symbol often irrespective of the language in which it is used).

5.8 Influence by Vedic-Hindu-Buddhist legacy

According to M.K. Agarwal (MKA) the origin of world civilization can be traced to the Sindhu and Sarasvati river valleys (located in present–day Pakistan) as early as 8000 BC. Here, innovation and originality in every aspect of human endeavor, from mathematics and science to art and sports, flourished. Yet the importance of this civilization, known as the Vedic period, has been deliberately downplayed.

Thoroughly researched and including an extensive bibliography, "From Bharata to India" rectifies this mistake in the perspective of world history and seeks to offer a comprehensive reference source. Author MKA shows how this early culture, where ideation (i.e., the innovative process of generating, developing, and communicating new ideas, where an idea is understood as a basic element of thought that can be either visual, concrete, or abstract. Ideation comprises all stages of a thought cycle, from creation, to development, to actualization. It is the essence of a design process, both in education and in practice) by enlightened philosopher Brahmin kings, brought material and spiritual wealth that was to remain unchallenged until the colonial era. This Vedic-Hindu-Buddhist legacy subsequently influenced peoples and paradigms around the globe, ushering in an era of peace and plenty thousands of years before the Europeans.

By using original sources in Sanskrit as well as regional literature, MKA compares corresponding situations in other civilizations within the context of their own literary traditions and records to prove that Bharata forms the basis of world civilization. This is in direct contrast to the "Greek or Arab miracle" hypothesis put forth by numerous scholars.

The first of two volumes in this series, "From Bharata to India" offers a fascinating, indepth glimpse into ancient India's contribution to the modern world.

5.9 Why base 10 number system survives and used by one and all

We have seen in the history of numbers the sexagesimal (base 60) system that was used in Mesopotamia by around 300 BC (without explicit use of zero). There are other countries where bases other than 10 have been used. However, it is base 10 that continues to be the sole base of the number system that is so well adapted with the whole of mankind today and will continue to remain so indefinitely. If the base is too large, such as 60, then comprehending the large number of symbols is difficult for a common man, although the physical size of the number (needing much less space in writing) could be small.

If, on the other hand, the base is small, such as 2 or 3, then once again comprehension is difficult since the physical size of the number is large (requiring much larger space in writing). So an optimum (not too large and not too small) base is desired taking into consideration the advantages (such as the number of fingers) extended to us by nature. The optimal base was thus found to be 10 and not (8 or 9 or 11 or 12). A human being has 10 fingers which have been used by people around the globe for counting. Also, base 10 (as it is not too large) is easily comprehended by a human being. Since we are accustomed so extensively that any other base will not be appealing to us, neither will it serve a better purpose in all contexts and environments taking into account our psychological bent.

For animals and birds about whom our knowledge is very limited, base 10 might not be so appropriate, nor are we overly concerned about them. In computer hardware, it is the binary number system that is used as in nature (silicon technology) there are only two very stable fast switching states. In computer software, we have the decimal (base 10) outputs as the interface between a computer and a user through an appropriate computer software/firmware.

5.10 Universally accepted zero is the zero viewed and used by Indians

The zero viewed and used by the Indians is a unique multipurpose mathematical object, that is, a number, a symbol, a direction separator, a magnitude, and a place-holder. This zero is a five-in-one operating perfectly well with a fully established positional number system. The Indian zero is now the universal zero and is time-invariant.

5.11 Psychological aspects of zero is distinct from those of nonzero numbers

No number such as 1, 2, 4, and 7 conjures up so much of psychological effect on a human being as does the number 0. The imagination of 0 as "nothing" when allowed to roam freely without any bound has the potential to induce a dreadful feeling of a universe in which one (he/she) is too tiny a part and is vanishing to a knot. To withstand this feeling, sufficient spiritual strength (possibly in terms of detachment) is called for. Such a feeling is completely absent with any other number.

As mentioned in Section 2, even the most well-known twentieth century spiritual giant Swami Vivekananda (SV), possessor of extraordinary mental capabilities, was horrified when his mentor and super-teacher—the most renowned nineteenth century mystic saint Sri Ramakrishna (SR)—touched him with his right foot. SV was then a young man of less than 19 years of age with an excellent physical build and strength. The mystic's touch produced a strange terrifying experience in him.

5.12 Computational zero versus absolute zero and error

Computational or, equivalently, numerical zero was never more prominent before than it is today. In the background of zero or, equivalently, absolute zero, also may be called the universal zero which is unique and context-independent, the computational zero is nonunique and context-dependent. The real, nondiscrete or, equivalently, continuous, quantity such as milk, height, and weight can never be exactly (error-freely) known. *It was never exactly known in the past, it is not exactly known in the present, and will never be known exactly in the future.*

A measuring device cannot usually measure any quantity with an accuracy greater than 0.005%. All that we can say is: the exact quantity lies in an interval depending on the accuracy associated with the concerned measuring device. We will discuss this aspect later in this chapter. The important requirement is to specify the narrowest possible proven interval or bounds so that the specified inexact quantity is acceptable in a given context. One can always specify a sufficiently large interval in which the exact quantity lies. But such a large interval has no practical/engineering utility and hence is meaningless/useless. Thus, associated with any real continuous quantity, there is always an error. While this error is never exactly known, its order is known.

5.13 Zero in natural mathematics, mathematics, and computational mathematics

We may divide mathematics into three categories—natural mathematics, computational or, equivalently, computer mathematics, and mathematics (employed by a mathematician). Natural mathematics is the mathematics used by nature all the time throughout the universe. This mathematics follows all the laws of nature (known or unknown to us), knows no error, represents all quantities exactly, almost all of which can never be captured by any means by any human being/living being, and uses infinite precision always for all computations. Computer mathematics, on the other hand, follows only known man-made rules, knows only errors, represents almost all quantities nonexactly, and uses finite precision always for all computations. Mathematics used by a mathematician are often expressed symbolically, tends/attempts to capture some of the aspects of natural mathematics, and uses the much fewer laws of nature known to human beings. While it can never capture error exactly, it symbolically provides bounds for an error; such bounds may or may not be useful in practice.

The computer arithmetic—a vital feature of computer mathematics—is essentially the IEEE 754 floating-point arithmetic. It often uses additional features, taking full advantage of binary representation in the hardware computer. The IEEE standard specifies three formats—single (32 bits), double (64 bits), and double-extended—of floating-point numbers. Each format can represent +0, −0, NaN (Not-a-Number), $\pm\infty$ (infinity), and its own set of finite real numbers all of the simple form $2^{k+1-N}n$ with two integers n (signed significant) and k (unbiased signed exponent) that run through two intervals determined from the format. Each of zero and infinity has two representations besides NaN (produced when, for instance, division by zero is encountered) in this representation.

Besides IEEE 754 which is a binary standard, there is the IEEE 854 standard that allows the radix 2 or 10. The radix 10 represents the conventional number system used and thoroughly understood by humans all over the globe and is specially suitable for calculators. Unlike IEEE 754, it does not specify how floating-point numbers are encoded into bits. The 854 standard specifies constraints on allowable values of the finite precision p for the single precision as well as for the double precision but it does not need a particular value for p.

The role of universal (absolute/unique) zero and that of numerical nonabsolute (nonunique) zero are well adapted in both IEEE 754 and IEEE 854 standards in a computer. This adaptation ensures a best computational accuracy (least computational error) for a specified precision. It can be seen that zeros—both absolute and numerical—play a vital role not only in improving accuracy and detecting illegal arithmetic operations (such as NaN's) but also in taking care of an overflow and/or an underflow appropriately. Consequently the possibility of undetected computational mistakes entering into the sequence of computations resulting in wrong/unacceptable outputs/results is eliminated/minimized unlike many pre-IEEE standards.

5.14 Is error in error-free computation zero?

Error-free computation for a problem/an algorithm can be carried out as long as the number of arithmetic (add, subtract, multiply, and divide) operations is finite and the input data are exactly (error-freely) representable in a finite precision computer. In general, the exact representation of the real world (continuous) input data is not possible. Consequently, there are errors in the input data. As a result, the output of error-free computation using any arithmetic (e.g., multiple modulus residue, p-adic, and rational) is not error-free. In fact the error gets increasingly *amplified* as the amount of arithmetic computation grows. While ideally we should get zero error in the output result, we simply do not get it.

However, in a hypothetical case where the physical input data are represented errorlessly in a computer, *only then we get zero-error results*. Such situations/problems in a physical world environment are too few compared to the number of problems that we face every day in practice. *Observe that exact error can never be computed; only an error-bound containing the exact error can be known/computed.*

5.15 Abbreviation involving letter O

It is sometimes convenient/confusion-free to write a small o for the letter O. For example, Ministry of Defence (or, Ministry Of Defence, upper case 'O' in 'Of' is normally not used though) is written as MoD rather than MD or MOD. Clearly, in the context, MoD is a better abbreviation than either (or better than even M.D.). The small o which looks very much like the symbol of zero is the distinct lower case version of the letter O and should not be confused/read as zero.

5.16 Is there anything beyond the fundamental particle considered as a building block of matter?

A physicist has always focused his mind in search of a building block of matter—a subatomic particle or subatomic elementary particle which is a *single entity and cannot be further subdivided*. A molecule followed by an atom followed by an electron. a proton, an antiproton, and a neutron each having a positive mass (very small though) have been discovered (initially as building blocks) to support a proposed atomic model.

The Rutherford-Bohr atomic model (1913) discussed in Chapter 2 was then accepted by physicists as the one in which particles such as electrons, protons, and neutrons are the indivisible building blocks. This model did not involve subatomic elementary particles such as Quark and Higgs boson discovered later over decades. These are increasingly tinier particles in terms of mass. *Can anything be formed just as an entire piece which when successively divided results in something having exactly zero mass or, equivalently, something which is nothing? That is, is there no indivisible building block of matter?*

Interestingly, successive division or bisection of anything can never become exactly zero unless we reach infinity of divisions/bisections. Reaching infinity is impossible. Hence reaching exact zero is impossible in this way. A simple example in physics is that of successive removal of gas (say, air) from a container. By any physical means, it is not possible to remove the gas completely and make the container totally free from gas (a perfect vacuum). Pondering over such a question could lead an intense thinker, at least in his imagination, to the state of exact zero through extrapolation.

Yet another example is the successive reduction of the temperature of a body to absolute zero (zero Kelvin or *absolute zero* or, equivalently, *ultimate zero* is the lower limit of the thermodynamic temperature scale, a state at which the enthalpy and entropy of a cooled ideal gas reaches its minimum value, taken as 0). The theoretical temperature is determined by extrapolating the ideal gas law; by international agreement, absolute zero is taken as −273.15° on the Celsius scale (International System of Units), which equates to −459.67° on the Fahrenheit scale. The corresponding Kelvin and Rankine temperature scales set their zero points at absolute zero by definition. In these scales, a zero became an *uncrossable barrier*: the coldest temperature possible. Absolute zero is the state where a container of gas has been drained of all its energy.

In Einstein's theory of relativity, a zero became a *black hole*, a monstrous star that swallows entire suns. In quantum mechanics, a zero is responsible for a bizarre source of energy—infinite and ubiquitous, present even in the deepest vacuum—and a phantom force exerted by nothing at all. One may come very close (within millidegrees) to absolute zero, but not exactly to absolute zero. Only through extrapolation (as stated above) we can get to know that there is absolute zero (at which the volume will be a numerical zero and not exact zero).

The mass of a particle (including a photon) at rest (invariant mass—same for all observers in all reference frames) and that at motion (relativistic mass—dependent on the velocity of the observer) are strictly different according to physics. In physics if the mass of a particle at rest is zero, its mass in motion is greater than zero according to the special theory of relativity. However, any particle that has a material existence, cannot have exactly zero mass; it will have (at least) a numerical zero mass at rest though. For a materials scientist or, equivalently, a physicist, such a state of matter (particle) of exact zero (rest) mass never existed and will never exist.

A rishi (spiritual scientist), who goes beyond matter and enters into the realm of nonmatter, that is, the spiritual world/state of super-consciousness, would be able to fathom the ultimate truth—this is how he gets his first-hand experience of exact zero and hence it is a perfect proof (much more intense and convincing than *any mathematical proof which is not totally free from fuzziness*) for him and such an experience is beyond the scope of physics as it cannot transcend matter. An ordinary/common man in his mundane approach may not grasp the experience. The same common man will certainly have the foregoing perfect experience when he focuses his mind intensely (without any flickering), although such a concentration/meditation may be initially difficult for most of us. It may be remarked that *soul is nothing but your consciousness*.

5.17 Epoch: origin of an era

An epoch is an instant in time chosen as the origin (starting point) of a particular era. The "epoch" then serves as a reference point from which time is measured. Time measurement units are counted from the epoch so that the date and time of events can be specified unambiguously. Events taking place before the epoch can be dated by counting negatively from the epoch, though in pragmatic periodization practice, epochs are defined for the past, and another epoch is used to start the next era, therefore serving as the ending of the older preceding era. The whole purpose and criteria of such definitions is to clarify and coordinate scholarship about a period, at times, across disciplines.

For instance, the epoch of the Hebrew or Jewish Calendar (introduced in the twelfth century) is October 7, 3761 BC, while the epoch of the Masonic calendar's Anno Lucis era is January 1, 4000 BC. There are several attributes associated with the term epoch such as geologic epoch (a span of time larger than an age but smaller than a period) and cosmologic epoch (a phase in the development of the universe since the Big Bang). As a matter of fact, a reference point in one or more dimensions is necessary with respect to which we describe the position or the time of a specific event/object. Without a reference point, it is not possible to specify anything. As anyone who has had to graph a triangle or a parabola knows Descartes' origin (0, 0) as the reference point in two dimensions prescribed by Rene Descartes (1596–1650), the founder of the Cartesian coordinate system.

5.18 Revelation through intense concentration

It may be remarked that revelation usually takes place through intense concentration as most of the important discoveries in sciences and engineering are the outcome of intense concentration/focus (in the calmest mind in which one can dive much deeper (into his/her mind) in search of the desired knowledge that happens to be the answer of the query) on the concerned subject matter/query (as against randomly generated thoughts arising in the ordinary/elementary mind). It may be stressed that *Mind is the reservoir of endless knowledge*. We can never create knowledge that does not exist at all or, equivalently, that does not exist in one's own mind. That is, all knowledge exists in everybody's mind; one just needs to mine the required knowledge out of the infinite ocean of knowledge residing in one's mind.

In contrast, the knowledge contained in all the computers (hundreds of millions of them) of the whole world is always finite and will continue to remain so through eternity, although growing every moment at an alarming rate with time; this will continue to ever remain ridiculously too negligible—a numerical zero—compared to that existing in one's mind! Here lies the vital difference between a living computer, that is, a living being (a human being, for instance) and a nonliving computer (a modern digital computer, for example), that is, a nonliving being.

The "Meerut incident" (Section 2.3.4) in the life of Swami Vivekananda and his attaining Nirvikalpa Samadhi (Section 2.3.3) constitute the extraordinary concentration as well as the ultimate concentration (no other state of mind exists beyond the Nirvikalpa Samadhi state anywhere in the universe in any living being), respectively.

5.19 Arabic inheritance of science from non-Arabic world and her contributions

The Arabic-language inheritance of science including mathematics was considerably Greek, a legacy of the rich Hellenistic tradition in Egypt and Syria. In 773 AD, at *al-Mansur's* (714–775 AD, the second Abbasid Caliph during 754–775 AD. He is generally regarded as the real founder of the Abbasid Caliphate) behest, translations were made of many ancient treatises including Greek, Latin, Indian, and others. This broad attitude (patronage of learning) of the then Arab rulers (with respect to pushing forward the frontiers of science) along with her further innovations contributed significantly to the world of sciences including astronomy and mathematics and enriched the world sciences, more specifically the European sciences during the second half of the first millennium AD.

5.20 Exact zero is usually unknown in physics

In the realm of continuous quantities in physics (and for that matter in any natural science such as chemistry and biology), exact zero is unknown. Any continuous quantity such as that of honey is never known exactly. We know at best bounds (enclosing a narrow interval depending on the error associated with the measuring device used) in which the exact quantity lies. As a matter of fact, *we know any continuous quantity only approximately (and never exactly) in all sciences and engineering*; exactness is completely unknown. *All the continuous quantities that we present in physics are invariably erroneous.*

Even if the quantity—such as the number of *red blood cells* in a sample of blood or, equivalently, red blood cells count—is not strictly continuous, it is often expressed approximately (since the number is too large), in terms of millions of blood cells in 1 mL^3 of blood in the foregoing count. Exact zero when used in the context of a continuous quantity, however, will simply imply the complete absence of the quantity.

5.21 Quantum, relativistic, and absolute zeros in physics are not exact zeros

In 1900, *Max Karl Ernst Ludwig Planck*, (1858–1947), a German theoretical physicist and originator of quantum theory, derived the formula for the energy of a single *energy radiator*, for example, a vibrating atomic unit:

$$\in = \frac{h\nu}{e^{\frac{h\nu}{kT}} - 1}$$

where the Planck's constant is $h = 6.62606957 \times 10^{-34}$ J.s (Joule.second), $\nu = 540 \times 10^{12}$ Hz (say) is the frequency of the associated electromagnetic wave, $k = 1.3806488(13) \times 10^{-23}$ J/K is Boltzmann's constant, and T is the absolute temperature.

Then in 1913, using this formula as a basis, Albert Einstein (1879–1955) and Otto Stern (1888–1969) published a paper in which they suggested for the first time the existence of a residual energy that all oscillators have at absolute zero. They called this *quantum zero* (residual) energy as *zero-point energy*. They carried out an analysis of the specific heat of hydrogen gas at low temperature, and concluded that the data are best represented if the vibrational energy is

$$\in = \frac{h\nu}{e^{\frac{h\nu}{kT}} - 1} + \frac{h\nu}{2}$$

According to this expression, an atomic system at absolute zero retains an energy of $0.5h\nu$ which is not exact zero.

Relativistic mass is the sum total quantity of energy in a body/object or a system of bodies (divided by c^2, where c is the velocity of light). As seen from the center of momentum frame (i.e., observer moving with the object), the relativistic mass is also the invariant mass (just as the relativistic energy of a single particle is the same as its rest energy, when seen from its rest frame). For other frames, the relativistic mass (of a body or system of bodies) includes a contribution from the "net" kinetic energy of the body (the kinetic energy of the center of mass of the body), and is larger the faster the body moves. Thus, unlike the invariant mass, the *relativistic mass* depends on the observer's frame of reference. However, for given single frames of reference and for isolated systems, the relativistic mass is also a conserved quantity.

Alternative theories of the photon include a term that behaves like a mass, and this gives rise to the very advanced idea of a "massive photon." If the rest mass of the photon were nonzero, the quantum electrodynamical theory would be in trouble basically through loss of gauge invariance, which would make it nonrenormalizable; also, charge conservation would no longer be absolutely guaranteed, as it is if photons have zero rest mass. But regardless of what any theory might predict, it is still necessary to check this prediction by doing an experiment.

However, it appears impossible to do any experiment that would establish the photon rest mass to be exactly zero. The best we can hope to do is place limits on it. A nonzero rest mass would introduce a small damping factor in the inverse square Coulomb law of electrostatic forces. That means the electrostatic force would be weaker over very large distances but not exact zero.

Both *quantum zero* and *relativistic zero* are bizarre and counterintuitive. The first describes that a vacuum (empty space) is not only nonempty but is also "full of force." The other says that the universe as we understand it could be populated by many so-called "*black holes*" (in mathematical lingo "*singularities*") within which the known laws of physics appear to break down. Like dividing by zero, attempts to unveil the mysteries of these physical zeros often bring about what appear to be paradoxes to the finite mind. And then there is the "most tantalizing" and "most unfathomable" zero of all in the universe: the birth of our own universe at the *"zeroth" hour of the "Big Bang."*

5.22 Zero or dot symbolizing beauty and eye of knowledge in Indian poetry and culture

In India, the use of zero and the place-value system has been a part of the way of thinking for so long that people have gone as far as to use their principal characteristics in a subtle and very poetic form in a variety of Sanskrit verse. As proof, here is a quotation from the poet *Biharilal* who, in his *Satsai*, a famous collection of poems, pays homage to a very beautiful woman in these terms: "The dot [she has] on her forehead increases her beauty tenfold, Just as zero dot (shunya–bindu) increases a number tenfold."

First of all, it should be remembered that the dot that the woman has on her forehead is none other than the *tilaka* (literally: sesame), a mark representing for the Hindus the third eye of Shiva, that is the eye of knowledge. While young girls put a black spot between their eyebrows by means of a nonindelible coloring matter, married women put a permanent red dot on their foreheads; it would seem then that the homage was being paid to a married woman. It is known that the dot (bindu) figures among the numerous numerical symbols with a value equal to zero, and is even used as one of the graphical representations of this concept. This is very clear allusion to the arithmetical operative property of zero in the place-value system, because if zero is added to the right of the representation of a given number, the value of the number is multiplied by 10.

Another quotation, taken this time from the *Vasavadatta* by the poet *Subandhu* (a long love story, written in an extremely elaborate language, swarming with word plays, implications and periphrases):

And at the moment of the rising of the Moon
With the darkness of the falling night,
It was as if, with folded hands
Like closed blue lotus blossoms,
The stars had begun straightway
To shine in the firmament (gagana)…
Like zeros in the form of dots (shunya–bindu),
Because of the emptiness (shunyata) of the samsara,
Disseminated in space (kha)
As if they had been [dispersed]
In the dark blue covering the skin of the Creator [=Brahma],
Who had calculated their sum total
By means of a piece of Moon in the guise of chalk.

Here too the metaphor used leaves the reader in no doubt; the void (shunya)—which is placed in relation to the emptiness (shunyata) of the cycle of rebirths (samsara)—is also implied in its representation in the form of a dot (shunya–bindu), as an operator in the art of written calculation. These concepts really had to have been part of the way of thinking for a long time for the subtleties used in this way to have been understood and appreciated by the wider public of the time.

5.23 Genius of Indian mathematical brains

Zero is not only inseparable from other nonzero negative and positive numbers but also pervades all of sciences and engineering. Today's mathematical knowledge is a gift from ancient India. Albert Einstein, one of the twentieth century's greatest brains with extraordinary depth of perception and far-sightedness coupled with superhuman critical analyzing power, said about India's contribution, "We owe a lot to the Indians, who taught us how to count through decimal system, without which no worthwhile scientific discovery could have been made."

5.23.1 Decimal system

The knowledge of nine numbers 1, 2, 3, 4, 5, 6, 7, 8, and 9 and the zero (0) which can be combined to form infinite mathematical expressions and measurements is considered the unique contribution of ancient Indian genius to world's progress.

During Vedic era, this decimal system was very much in vogue in India. The numerical values in a sequence such as *eka, dasa, sata, sahasra, ayuta, laksha, niyuta, koti, arbud, vrinda, kharav, nikharav, shankha, padma, sagar, antya, madhya, ..., and parardha* are described in the 2nd mantra of Chapter 17 of Yajur Veda Samhita.

"Lalita Vistara" (first century BC), a Buddhistic text describes up to 10^{53} and named that numerical value as *Talakslma*. *Anuyogadwara*, a Jain text describes numbers up to 10^{140}.

The ancient Greeks gave the biggest numerical value called *myriad* which is 10^4, that is, 10,000 while the biggest Roman numerals were 10^3, that is, 1000 only.

The numbers 0 to 9 were first adopted by Arabs from India and spread to Europe. Today we call these numerals as Indo–Arab numerals.

5.23.2 Glory of zero

The complete mathematical knowledge becomes "knot or zero" without India's richest zero which is used by Indians both as a mathematical expression and as a philosophical/spiritual concept. Vedas, Puranas, Upanishads, and many other Indian classical texts had dealt with zeros in many ways.

Pingala (about 500 BC) in his *Vedagana* text *ChandasSastra* (A Guide to Study Vedic Prosody), while explaining Gayatri Chandas mentions zero.

Gayatre sadsamkhyamardhes panite dvyanke avasista srayastesu
Rupamapaniya dvyankashah sunyam sthapyam!!

In mathematics, usage of negative numbers came into existence because of zero's invention. In Isavasya Upanishad, in the Shanti mantra, the following verse describes the philosophy of zero:

Purna madah purna midham purnat purna mudacyate
Purnasya purnamadaya purnameva vasisyate

"From zero or completeness everything came and into zero or completeness everything merges, zero or completeness alone exists." In Sanskrit, "purnam" is used to denote "zero" or "completeness."

The Jain text *Suryapragnapti* (400 BC) classifies the numbers into three categories and describes five types of infinity.

Add, subtract, multiply, divide, squaring, square-rooting, and cube-rooting operations were known to Indians and can be found in most of the mathematical texts of India. In the text *Ganita Sara Samgraha* of Mahaviracharya (850 CE) *Sridharacharya* explains LCM, zero, finding square-roots, and solving quadratic equation among others. In his text *Pathiganitam*, he describes several arithmetical problems such as concepts of calculating simple interest, compound interest, problems on time and distance, and time taken for filling the water. In Bakhshali Manuscript (about 200 BC), one can find negative numbers, fractions, (sequences of) arithmetic and geometric progressions, etc.

Geometry, an important branch of mathematics originated in India. The word *Jyamiti* is a Sanskrit word meaning measuring the earth. "Jya" in Sanskrit means earth, "miti" means measurement. Jyamiti or Geometry means measuring the earth. *Kalpa* sastra, a part of Vedangas consists of "Sulba Sutras," which explains the techniques of constructing *yajna vedicas* (vedic sacrificial altars and platforms). From these verses (sutras), Geometry evolved. Today what we call *Pythagoras Theorem* is a mere repetition of what has been said in *Baudhayana "Sulba Sutras"* written around 600 years before Pythagoras:

दीर्घ चतुरसस्य अक्ष्णया रज्जुः पार्श्वमानि तिर्यक् मानि च ।

यत् पृथग्भु ते कुरुतः तद् उभयं करोति ॥

Dirghacaturasasyaksnya rajjuh parsvamani tiryak mani ca |
yatpthagbhu te kurutah tadubhayam karoti ||

Of a rectangle, when the sides of the squares obtained separately by stretching a rope along the sides (length and breadth) and diagonal of the rectangle, the areas (of the squares) produced separately by the two sides of the rectangle equal the area produced by the diagonal. Equivalently, in a right-angled triangle, the square of the diagonal on the hypotenuse is equal to the sum of squares of other two sides (Baudhayana Sulba Sutras Chapters 1–12 sloka).

5.23.3 Value of pi

Pi, the circumference divided by the diameter of any circle, that had attracted the attention of almost all mathematicians—ancient and modern, Indian and western—is a universal constant. Aryabhatta (born 2765 BC) had calculated the value of Pi as 3.1416 which is correct up to five significant digits where the number of significant digits = lower integral part of

$$\log_{10}(1/\text{relative error}),$$

where

$$\text{relative error} = |\,3.141592653589793 - 3.1416\,| \,/\, |\,3.141592653589793\,|,$$

the first number is the higher order accurate value of Pi or, equivalently, sufficiently more accurate value of Pi that is taken to replace the exact value of Pi (the exact value of the irrational number Pi or any other irrational number such as $2^{0.5}$ is never known in a finite number of digits in any meaningful number system, it exists in its exact form physically though). As a matter of fact any continuous quantity (and even a noncontinuous quantity having extremely large number of individual entities such as the number of red blood cells in 1 mL^3) can never be exactly representable. Further, Bhaskaracharya (working 486 AD), Mahaviracharya, also known as Mahavira (817–875 AD, not to be confused with Vardhamana Mahavira of nineteenth century BC), Nilakanthan Somayaji (1444–1544 AD), and Srinivasa Ramanujan (1887–1920 AD), among many others, calculated the value of Pi.

Brahmagupta in his text Brahmasputa Siddhanta (Chapter 12, Verse 28) describes the mathematical methods to find the lengths of diagonals of a rectangle that is embedded in a circle.

Bhaskaracharya in his book *Leelavati* describes cyclic quadrilaterals, cyclic pentagons, cyclic hexagons, and cyclic octagons, and further postulates that sides of quadrilaterals and the diameter of the circle that is circumscribing them shall be in a constant ratio.

Aryabhatta in his text *Aryabhattiyam* gives the formula for calculation of area A of a triangle as the product of half the base b of the triangle and the height h of the triangle, that is, A=0.5bh:

त्रिभुजस्य फलशरीरम् समतलकोटि भूज अर्ध समवर्गः ।

Tribhujasya phala sariram samadalakoti bhujardha samvargah |

The product of the base length and half the perpendicular on the base determinines the area of a triangle.

5.23.4 Trigonometry

Trigonometry is a gift from ancient India to the mathematical world. The concept of sine and cosine had been evolved by Indian mathematicians. Bhaskaracharya had postulated various trigonometric principles and equations in his text *Leelavati*. Varahamihira, Brahmagupta, and Lalla and other Indian mathematicians had given various trigonometric formulas. The Kerala mathematician Madhava, in his book *Karanapaddhati*, dealt extensively with trigonometric formulas and functions.

5.23.5 Calculus

What we call today Calculus was known by the ancient Indians as *Kalana Ganana Sastra*. Long before Newton and Leibnitz had made use of it, Aryabhatta and Bhaskaracharya had dealt with this branch of mathematics in their astronomical calculations. Bhaskaracharya in his text *Siddhanta Siromani* (Chapter 4, Graha Ganita) deals with the concept of differentiation and its application by considering the temporal positions of various planets. Aryabhatta had pioneered the method of calculating the temporal positions of various planets and had introduced to the world the knowledge of calculus. Brahmagupta and Madhava (1340–1425 AD) had developed this branch of mathematics by introducing integral calculus.

5.23.6 Algebra

Algebra, a branch of mathematics, is also an Indian invention. During the ninth century AD, Arabs adopted it and from them it spread to the other parts of the world. Indian seers of yore such as *Āpastamba* (600 BC), Baudhayana (800 BCE), and *Kātyāyana*, (third century BC)—a Sanskrit grammarian, mathematician, and Vedic priest—in their Kalpasūtras had introduced the "unknown" value/variable in their mathematical expressions. The *Dharmasutra* of *Āpastamba* forms a part of his larger Kalpasūtra. It contains thirty *praśnas*, (literally meaning "questions"). The subjects of this Dharmasūtra are well organized and preserved in good condition. Afterwards, Bhaskaracharya, Madhava, and others developed various algebraic formulas, equations, and functions.

Bhaskaracharya calls this subject *Ayaktaganita* or *Bijaganita* which is Algebra. He had said that *Vyakthaganita* (Arithmetic) leads to *Ayaktaganita*. In his book Leelavati, he deals with *Vyakthaganita* before dealing with *Ayaktaganita*.

Indian mathematical genius is evident from the seers of Vedic times to twentieth century Srinivasa Ramanujan. Today, what we call computer language (Backus–Naur Form, or Backus Normal Form) is a replication of Panini's (about 500 BC) grammar rules. It may be observed that all the foregoing subject areas would not have flourished or even existed if zero in a number system had not have been given the most respected position.

5.24 Thoughtful comments/convictions

The feeling and realization concerning zero by an intense thinker (mathematicians, astronomers, philosophers, and others) provide us with the food for thought and consequent realization/revelation when we dive deep (or focus our mind) into the thinking process on zero. Following are some of the realizations by celebrities—explicitly known and unknown, which provide enough information to seriously ponder over assimilating the character (dreadful on one hand and serene on the other) and perpetual/eternal importance of zero.

C. Seife: Zero is powerful because it is infinity's twin. They are equal and opposite, yin and yang. They are equally paradoxical and troubling.

A.N. Whitehead: The point about zero is that we do not need to use it in the operations of daily life. No one goes to buy zero fish. It is in a way the most civilized of all the cardinals, and its use is only forced on us by the needs of cultivated modes of thought.

T. Danzig: In the history of culture the discovery of zero will always stand out as one of the greatest single achievements of the human race.

C. Seife: The Babylonians invented it, the Greeks banned it, the Hindus worshiped it, and the Church used it to fend off heretics. For centuries, the power of zero savored of the demonic; once harnessed, it became the most important tool in mathematics. Zero follows this number from its birth as an Eastern philosophical concept to its struggle for acceptance in Europe and its apotheosis as the mystery of the black hole. Today, zero lies at the heart of one of the biggest scientific controversies of all time, the quest for the theory of everything. Elegant, witty, and enlightening. Zero is a compelling look at the strangest number in the universe and one of the greatest paradoxes of human thought.

C. Seife: Zero finally appeared in the East, in the Fertile Crescent of present–day Iraq.

C. Seife: By around 300 BC the Babylonians had started using two slanted wedges to represent an empty space.

C. Seife: The Mayans had a zero in their counting system, so they did the obvious thing: they started numbering days with the number zero.

C. Seife: If you want only divide by zero, you can destroy the entire foundation of logic and mathematics.

Lucretius: Nothing can be created from nothing.

C. Seife: Around 500 BC the place-holder zero began to appear in Babylonian writings; it naturally spread to the Greek astronomical community. During the peak of ancient astronomy, Greek astronomical tables regularly employed zero; its symbol was the lowercase omicron, ø, which looks very much like our modern–day zero, though it's probably a coincidence. (Perhaps the use of omicron came from the first letter of the Greek word for nothing ouden.)

The calculus: Brings the concept of limit zero.

The Rig Veda: In the earliest age of the gods, existence was born from nonexistence.

C. Seife: Indian mathematicians did more than simply accept zero. They transformed it, changing its role from mere place-holder to number. This reincarnation was what gave zero its power.

Archimedes rejected zero, which is the bridge between the realms of the finite and the infinite, a bridge that is absolutely necessary for calculus and higher mathematics.

Holy Koran or *Quran*—the central religious text of Islam, the greatest literary work in classical Arabic, believed to have been revealed by God to Prophet Muhammad (570–632), whose full name is Abū al-Qāsim *Muḥammad* ibn Abd Allāh ibn Abd al-Muṭṭalib ibn Hāshim, through the angel Gabriel gradually over 23 years from December 22, 609 CE (when Muhammad was 40) to 632 CE, the year of his death—Does man forget that we created him out of the void?

5.25 Counting from zero? John Conway and Richard Guy

Waclaw Sierpinski, the great Polish mathematician … was worried that he'd lost one piece of his luggage. "No, dear!" said his wife. "All six pieces are here." "That can't be true," said Sierpinski, "I've counted them several times, zero, one, two, three, four, five." We may recall, in this context, Section 4.2.2 ("Zero is still causing problems!") and Section 4.3 (Y2K problem).

5.26 Root of the word zero

The Indian name for zero was sunya, meaning "empty," which the Arabs turned into sifr. When some Western scholars described the new number to their colleagues, they turned sifr into a Latin-sounding word, yielding zephirus, which is the root of our word zero. Other Western mathematicians did not change the word so heavily and called zero cifra, which became cipher. Zero was so important to the new set of numbers that people started calling all numbers ciphers, which gave the French their term chiffre, digit.

5.27 Zero and infinitely small in mathematics, physics, and beyond

Zero appeared in the middle of every Renaissance painting where a vanishing point links zero and infinity. Zero on a real line (one-dimensional), that on a plane (two-dimensional), and that that on a hyperplane (multidimensional) may be viewed as 1-, 2-, and multidimensional zeros. A zero-dimensional zero may be considered as an extension/extrapolation of multidimensional zero. The *electron* is a zero-dimensional object, and its very zero-like nature ensures that scientists don't even know the ecotones (an ecotone is a transition area between two biomes) mass or charge. Also, *black holes* are zero-dimensional. A black hole is a zero in the equations of general relativity; the energy of the vacuum is a zero in the mathematics of quantum theory.

While *Evangelista Torricelli* (1608–1647), an Italian physicist and mathematician, best known for his invention of the barometer, created the first vacuum (numerical void, not exact void since an exact void cannot be created), *Blaise Pascal* (1623–1662), a French mathematician, physicist, inventor, writer, and Christian philosopher investigated the nature of a void.

The term "*Infinitely small*" came to be used (specifically in mathematics) as an important technical term and continues to remain in use. Zero and Infinity, on the other hand, are simply viewed opposite poles on the Riemann sphere and the infinity of rationales is nothing more than a zero (occupy space on the real line).

5.27.1 Zero space

In mathematics, a *zero-dimensional topological space* is a topological space that has dimension zero with respect to one of several inequivalent notions of assigning a dimension to a given topological space. Specifically: a topological space is zero-dimensional with respect to the Lebesgue covering dimension if every open cover of the space has a refinement which is a cover of the space by open sets such that any point in the space is contained in exactly one open set of this refinement. On the other hand, a topological space is zero-dimensional with respect to the small inductive dimension if it has a base consisting of clopen sets. These two notions agree for separable, metrisable spaces.

A zero-dimensional Hausdorff space is necessarily totally disconnected, but the converse fails. However a locally compact Hausdorff space is zero-dimensional if and only if it is totally disconnected.

Zero-dimensional Polish spaces are a particularly convenient setting for descriptive set theory. Examples of such spaces include the Cantor space and Baire space.

5.27.2 The quantum universe: a zero-point fluctuation?

In the past, scientists were more inclined to believe in a static universe. But by the early-twentieth century they learnt that the universe was not static, and asked the questions "Where were we from?" and "Where are we heading to?" When these questions are posed with general theory of relativity and the theory of quantum mechanics—as appeared relevant at the Third Loyola Conference on Quantum Theory and Gravitation, held at Loyola University in New Orleans in 1979—some ideas were born.

Alexander Vilenkin (born 1949), a theoretical physicist of Tufts University in Medford, Massachusetts, proposes to demonstrate that the universe was created from nothing. As he points out, "the idea is very old in the context of theology." Edward P. Tryon, an American professor of physics at Hunter College (of the City University of New York) in Manhattan, who specialized in theoretical quark models, theoretical general relativity, and cosmology proposed that the universe is a large-scale vacuum energy fluctuation, also called the zero-energy universe hypothesis. He has been quoted as saying, "the universe is simply one of those things, that happens from time to time."

What Tryon noticed was that *over the whole universe many of the conserved quantities of physics add up to zero.* Conservation laws—which state that the total amount of some quantity in a system undisturbed from outside does not change—are essential to physical analysis. Quantities such as energy and momentum are conserved. On the subatomic level many of the quantities, collectively known as quantum numbers (quantum numbers describe values of conserved quantities in the dynamics of a quantum system), that differentiate one kind of particle from another are conserved. As it happens, many of those that have positive and negative aspects, like electric charge, add up to zero over the whole universe. Vilenkin points out that *cosmologists can make others come to zero also by choosing boundary conditions appropriately.*

All this implies that the universe could be a "quantum mechanical fluctuation." In quantum mechanics, zero does not always remain zero. In a balanced situation, as this kind of thinking postulates for the universe, the positives and negatives can separate enough for some physical processes to occur for a very short time. Quantum mechanical fluctuations generally last for so fleeting a time that the measuring devices that we have so far created cannot be sure they (the fluctuations) existed at all. If the universe is one, and we are living inside it, it is an interesting and paradoxical place to be. The fluctuation may last eons (an eon denotes an indefinitely long period of time. It is the largest division of geologic time comprising two or more eras or 1 billion years or more) for us who are inside the universe (assuming the universe is one), but hardly any time at all from an outside point of view, about which we are yet comprehensively ignorant.

Vilenkin stresses that until recently the Big Bang theory concerning the universe's growth clashed with such ideas of a zero-point beginning. The theory never precisely stated how large the universe was when it started to expand/grow, but apparently the universe had to be larger than a fluctuation could be to reach its current size with a steady rate of growth. Now, however, cosmologists have inflationary growth scenarios, such as those championed by Alan H. Guth (born 1947) of Massachusetts Institute of Technology. Guth has also pointed out the possibility of creation ex nihilo. These theories postulate a period of very rapid expansion early in history, and so can accommodate the beginning that Vilenkin desires. That beginning involves another paradox of quantum mechanics: tunneling. Tunneling is a phenomenon frequently encountered in electronic circuitry—in Josephson junctions, for example. If there is an insulating gap in an electric circuit that represents an energy barrier greater than the energy possessed by moving electrons, current will not flow—according to the classical theory of electric circuits. "The universe arises by quantum tunneling from nothing—a state with no classical space-time."

However, if we consider the system quantum mechanically then we need to represent the electrons with wave equations that, among other things, enable physicists to compute the probability for electrons to be in one place or another. It turns out that the wave equations give a certain probability for electrons to be on the opposite side of the barrier, and in practice they turn out to be there. Nevertheless a certain current does flow.

To transcend from the spaceless, timeless state of nothingness to a state of somethingness in which both space and time as well as matter can exist needs the passage of a similar barrier. Vilenkin may write down a wave equation for the universe and "calculate the tunneling probability—whatever it means." The meaning of the probability is not all that clear, because in this instance Vilenkin is dealing not with the statistics of billions of electrons in a circuit, but solely with the one and only universe we know of. The philosophical foundations of quantum mechanics can get a bit self-contradictory when dealing with "one and only" universe.

Whatever probability may imply in the case of the one and only universe, Vilenkin may compute from the equation that the universe began as if it had come through such a tunnel, nucleating as a tiny bubble in which space and time existed and matter could be produced and even provide a formula to determine the bubble size, based on the density of the universe and the time since it came out of the tunnel. The bubble expands from there. He says that this universe is homogeneous and isotropic—two characteristics we observe in the universe today. Further, it is a closed universe. If the universe is a quantum mechanical fluctuation, it eventually has to relax back to the state of zero-oriented nothingness that it started from.

Physics limits itself to causes and effects only related to matter and not outside it. However, the Loyola discussion dealt with what could be the very first material cause and the very last material effects, and the participants had a tough time keeping God/nonmatter/spirit out of it. "Why is the universe so large?" asked Don N. Page. (Don N. Page is a Canadian theoretical physicist at the University of Alberta, Canada. His work focuses on quantum cosmology and theoretical gravitational physics, and he is noted for being a doctoral student of the eminent Prof. Stephen William Hawking (Stephen Hawking, born 1942, is an English theoretical physicist, cosmologist, author and Director of Research at the Centre for Theoretical Cosmology within the University of Cambridge), in addition to publishing several journal articles with him, he is an Evangelical Christian.) "To say that God created it is outside of physics." Spontaneously Page quoted the Bible about the initial condition of the universe: "without form, and void."

Normally to respond to the problem/question posed by Page is to build an appropriate mathematical model of the growth of the universe and solve it. Such a model, as Page pointed out, consists of three parts: (i) the physical quantities that vary, (ii) dynamical equations that describe the variations of the physical quantities and their mutual relations, and (iii) the boundary conditions, the special qualities and beginning and end values that apply to the particular case.

Arising out of the geometry of 3-D space and the matter field, the nature and constitution of matter in a specified space are the concerned physical variables when the universe is considered as one in its totality. The geometry could be (i) positively curved, that is, parallel lines eventually converge/meet by one or another amount, (ii) flat, that is, parallel lines remain equidistant throughout until infinity, or (iii) negatively curved, that is, parallel lines eventually diverge. Each combination of geometry and matter field comprises a distinct universe. Thus one can visualize an indefinitely large number of combinations and hence indefinitely large number of universes that could be imagined to exist parallel to one another.

To sort out this multiplicity, John Archibald Wheeler (John Wheeler, 1911–2008, was an American theoretical physicist who was largely responsible for reviving interest in general relativity in the United States after World War II. For most of his career, Wheeler was a professor at Princeton University, and was influential in mentoring a generation of physicists who made notable contributions to quantum mechanics and gravitation) introduced, about a couple of decades ago, the notion of a superspace, a hypothetical space in which each possible combination of geometry and matter field is represented by a single point. Page and Hawking do much of their mathematical operation in this Wheeler superspace. The dynamical equations, that is, the Hamilton's equation and the Schrödinger's equation. are two workhorses of theoretical physics, but the equations themselves will not suffice to explain the state of the real universe (the only one, we are certain, that exists) which we are in. It is a unique universe with a very special set of boundary conditions that theorists must consider. Our universe is homogeneous on the large scale, isotropic, seemingly flat or approximately so, and it has, according to Page, "a very strong arrow of time." The last point means that neither we nor any macroscopic method known to us can go backward in time. Page estimates that the chances of "the creator sticking in a pin" and pulling out just this combination of qualities that make such a unique universe are far beyond the astronomical figure, that is, 1 in (10,000,000,000) $\times 10^{124}$.

In the quest of a final boundary condition, the question that arises is: Does the universe have a geometric boundary? That is, if we go sufficiently far, will we fall off the edge? Hawking responded in the negative: "What could be more special than that the universe has no boundary?"

In mathematics, and particularly in set theory and the foundations of mathematics, a universe is a class that contains (as elements) all the entities one wishes to consider in a given situation. In set theory, a universal set is a set which contains all objects (all-inclusive), including itself. In set theory, as usually formulated, the conception of a universal set leads to Russell's paradox (according to simple set theory, any definable collection is a set. Let R be the set of all sets that are not members of themselves. If R is not a member of itself, then its definition dictates that it must contain itself, and if it contains itself, then it contradicts its own definition as the set of all sets that are not members of themselves. This contradiction is Russell's paradox) and is consequently not allowed. However, some nonstandard variants of set theory include a universal set.

The presence of Schrödinger's equation indicates that Hawking's and Page's work also includes a wave equation for the entire universe or, equivalently, a quantization of the whole universe. Deriving a wave equation for something so macroscopic as the entire universe leads to several issues regarding the meaning of quantum mechanics. Page got into a pointed discussion of these issues with *Eugene Paul Wigner* (Eugene Wigner, 1902–1995, was a Hungarian American theoretical physicist and mathematician. He received a share of the Nobel Prize in Physics in 1963 "for his contributions to the theory of the atomic nucleus and the elementary particles, particularly through the discovery and application of fundamental symmetry principles."

Near the end of his life, Wigner's thoughts turned more philosophical. In his memoirs, Wigner said: "The full meaning of life, the collective meaning of all human desires, is fundamentally a mystery beyond our grasp. As a young man, I chafed at this state of affairs. But by now I have made peace with it. I even feel a certain honor to be associated with such a mystery." He became interested in the Vedanta philosophy of Hinduism, particularly its ideas of the universe as an all-pervading consciousness. In his collection of essays *Symmetries and Reflections—Scientific Essays*, he commented "It was not possible to formulate the laws (of quantum theory) in a fully consistent way without reference to consciousness."

Wigner also conceived the "Wigner's friend" thought experiment in physics, which is an extension of the "Schrödinger's cat" thought experiment. The *Wigner's friend* experiment asks the question: "At what stage does a 'measurement' take place?" Wigner designed the experiment to highlight how he believed that consciousness is necessary to the quantum-mechanical measurement processes.

A quantum mechanical wave equation can be interpreted as a predictor of probabilities. Very often such an equation represents what physicists call "a superposition of states." That is, it involves two or more states of being for a given system, say A and B. In dealing with a large number of similar microscopic objects such as billions of electrons, the physicist can say that a certain percentage of them are in state A and the remaining percentage in state B. If, instead of billions of microscopic objects, say, electrons, only one electron is involved, one can imagine that it is somehow alternately in each of the states. As it is not possible to measure the state or follow the action of a single electron, the physicist may get away with the apparent self-contradiction: Nobody may know exactly what is going on. With macroscopic objects, on the other hand, which can be measured reasonably accurately, and whose actions can be followed individually, the contradiction does hurt.

The cat analogy given by Erwin Schrödinger, one of the founders of quantum mechanics, does not appear to be appropriate. The wave equation for this cat is a superposition of two states, that is, live state and dead state. This is impossible. The cat is either alive or dead; it cannot be both at the same time, nor can it be alive at one instant, dead at the next and again alive at the following instant.

Wigner who in the paper presented at the conference raised the issue of the philosophical bases of quantum mechanics: "What happens when you make a measurement?" Arthur Komar (Arthur B. Komar, 1931–2011, of Yeshiva University was a theoretical physicist specializing in general relativity and the search for quantum gravity. He made a significant contribution to physics as an educator, research scientist, and administrator. He had wide interests in numerous other subjects) seconded Wigner, saying: "I have 100 percent probability [after making a measurement]; what is my wave function?" and "When you make a measurement the wave function collapses." That is, if one knows something for certain (that the cat is alive, perhaps), the wave function becomes meaningless.

Page replied that one cannot test the absolute probability of anything, and he quotes Hawking: "When I hear of Schrödinger's cat, I reach for my gun." Somewhat less categorically, Page stressed on relative probabilities: If one measures something to be A, he puts that in his memory bank and goes on to ask the wave equation, given A, what is the relative probability of B? Page feels that this is a meaningful way to proceed, although others (questioners) seem to remain unconvinced.

However, Page derived, proceeding in this way, a model that provided the universe the proper expansion, including the inflationary period at the very beginning that was necessary to get the universe to the correct present size, and did it without the fine tuning of conditions required by some other inflationary models. Page's model did not appear to explain the specific boundary conditions of our universe, nevertheless, he said, "A number of things have been done with this. This proposal does agree with some observations."

As to the universe, cosmologists are more concerned with the beginning and the middle; they appear less keen in considering the end of the universe. Probably this is due to the fact that they do not know which end it will be. It is widely believed that the universe started very small and expanded. The end depends on whether the universe is open or closed. The universe as we have it appears to be nearly flat (which for the discussion that follows is a special case of an open universe). However, if there is a slight curvature, it could be in the direction of a closed universe. Frank Jennings Tipler (Frank Tipler, born 1947, is a mathematical physicist and cosmologist, holding a joint appointment in the Departments of Mathematics and Physics at Tulane University in New Orleans) presented a scenario for each case as follows.

An open universe will expand unendingly, and things will gradually run down like a clock that nobody winds. A closed universe, on the other hand, will reach some maximum size and then eventually collapse, coming to what Tipler calls "*a crunch singularity*," a state in which *temperature and density become infinite and the radius of the cosmos (universe) is zero*. Both scenarios leave no hope for the endless survival of humans made of flesh and blood. However, Tipler holds out a hope for the survival of some kind of (living) being capable of storing and processing information.

Tipler's open and flat universe scenario starts with the sun deserting the main sequence of stellar evolution in a billion years and becoming unreliable as a steady source of energy. In a trillion years stars will cease to be born. Stars then start to cool off until, after 10^{15} years, and we will have dead planets unattached with dead stars.

As the protons in dead stars decay during the period from 10^{31} to 10^{34} years, the dead stars evaporate. Black holes of around the sun's mass then decay by giving off radiation in a manner that Hawking proposed a few years earlier. This is over by 10^{54} years. By 10^{71} years most of the electrons and positrons left over from the decays form positronium (positronium is a system consisting of an electron and its antiparticle, a positron, bound together into an *exotic atom*, specifically an *onium*).

The system is unstable: the two particles annihilate each other to produce two gamma-ray photons after an average lifetime of 125 ps (1 ps = one trillionth of a second = 10^{-12} s = 1/1,000,000,000,000 s) or three gamma-ray photons after 142 ns (1 ns = one billionth of a second = 10^{-9} s = $^{1}/1,000,000,000$ s) in vacuum, depending on the relative spin states of the positron and electron. The orbit of the two particles and the set of energy levels are similar to that of the hydrogen atom (electron and proton), a quasiatom in which an electron and a positron are bound together and orbit around each other. By 10^{102} years even the most massive black holes, those with masses as large as superclusters of galaxies, have evaporated into Hawking radiation.

The fates of open and flat universes have slight difference in the last eons. The positronium in a flat universe will decay as electrons and positrons meet each other and annihilate each other into photons or light particles. By 10^{116} years only photons and a few residual protons are all that remain. The universe is now so spread out that these protons are the equivalent of black holes, and they decay by Hawking radiation. By 10^{128} years there remains only silence, cold and some incredibly dim and spread-out light.

In the open universe, on the other hand, some of the positronium would remain. The average distance between electron and positron in these relict (a relict is a surviving remnant of a natural phenomenon) quasiatoms (quasiatoms are collections of atomic or subatomic particles that when undergoing a collision that briefly appear to have the same characteristics as a (larger) atom. This can occur when the nuclei of the two sets of particles colliding become much closer to each other than they are to their constituent electrons. The combined nuclei then exhibit the same 1/r force as a single nucleus, where 1/r implies inverse distance law) would be 100 times the present size of the universe, and their orbital speed a micron per century.

In the closed universe, the cosmos must first reach its maximum size—we know it is still expanding now—and then start to collapse. Hence the time scale is uncertain. When the universe gets down to 1/100 of its present size and a temperature of 100 K (compared to the present 3 K), galaxies merge. At 1/1,000 (a thousandth) of present size, the sky becomes as bright as the sun and the temperature is 3000 K. At 1/1000000 (a millionth) of present size and 3 million K, the sky is as hot as the cores of stars. At 1/1000000000 (a billionth) of its present size (which equals about 10 light-years for the diameter of the universe) and 30 million K, atomic nuclei are dissociated into neutrons and protons. Finally, at 1/10000000000 (a ten-billionth) of its present size (or about a hundredth of a light-year) and 30 trillion K, neutrons and protons disintegrate into quarks.

"It looks bad for mankind," says Tipler. "We're getting clobbered." But such a gloomy assessment, he says, is based "only on the idea that intelligence can be coded only in *Homo sapiens*."

Suppose some intelligent machine existed that could withstand the physical conditions of the far future. It would have to find the energy to process information and have a mechanism for storing the information. Tipler finds such conditions in the closed universe, and he says, "If life—when defined as information processing and storage—exists forever, the universe must be closed." Borrowing an idea from Pierre Teilhard de Chardin (de Chardin, 1881–1955, was a French philosopher and Jesuit priest who, trained as a paleontologist and geologist, took part in the discovery of *Peking Man*. He conceived the idea of the Omega Point (a maximum level of complexity and consciousness towards which he believed the universe was evolving) and developed Vladimir Vernadsky's concept of noosphere to denote the "sphere of human thought"), Tipler predicts that "the universe must have one point singularity—a point toward which everything converges—in the customary cosmology. Call it an omega point."

5.27.3 Dark matter: still in the dark in our quest for the origin of the universe

Dark matter (DM) is the greatest scientific mystery known—an invisible substance thought to constitute up to (or more than) 83% of all matter in the universe. It remains so mysterious that scientists are still uncertain as to whether DM or nonradiating matter is made of microscopic particles or far larger objects. The consensus right now is that DM consists of a new type of particle, one that interacts very weakly at best with all the known forces of the universe except gravity. As such, DM is invisible and mostly intangible, with its presence only detectable via the gravitational pull it exerts.

Past research has discovered super-massive black holes millions to billions of times the mass of the sun in the heart of galaxies, but these are only detectable because they are so large, conspicuously disrupting matter around them. In theory, much smaller black holes could have formed in the early universe. These so-called primordial black holes would be far more difficult to detect, and they could potentially exist in large enough numbers to make up all DM.

It is the second decade of the twenty-first century. We have yet to fully enlighten ourselves regarding the DM that is believed to dominate the masses of all galaxies. The consensus of cosmologists now is that it is a new kind of particle which does not couple with radiation. However, the evidence for the existence of such DM goes back to the early/mid-1930s when the Swiss-American astronomer Fritz Zwicky was attempting to estimate the mass of large clusters of galaxies.

A large galaxy such as our Milky Way is a gravitationally bound large cosmic structure consisting of some hundred billion stars. There are many galaxies (smaller than the Milky Way) having a billion (or less) stars. Just as a large number of stars clusters to constitute a galaxy, galaxies themselves group to form large galactic clusters which may have 30,000 or more galaxies. Further, clusters themselves can group to form superclusters. The motion of the Sun around the galaxy enables us to estimate the total mass of the galaxy just as the earth's orbital velocity and distance from the Sun enables us to estimate the mass of the sun using Newton's gravitational force laws and Kepler's laws. In the same way, the motion of the galaxies in a cluster of galaxies and the size of the clusters allows us to deduce the local mass (called dynamical mass) of the cluster. This is how Zwicky and others estimated the mass of several rich galaxy clusters.

There is yet another way of estimating the mass of the cluster. By knowing the mass of a typical galaxy and its luminosity (i.e., the power in watts emitted due to all the stars, nebulae) and by estimating the luminosity of the galactic cluster and comparing it with an individual galaxy, we can estimate the cluster mass. Thus if it is a million times brighter than a typical galaxy, then the mass must be a million times that of the galaxy. The mass measured in this way is known as the *luminous mass*.

Zwicky's observations were confirmed later by other astronomers. It is now accepted as an established paradigm although he overestimated the amount of DM. Indeed as much as 90% of the galaxy mass is due to DM. Most of the light from a galaxy comes from central regions, where most of the stars are concentrated. If most of the mass is also in the central part, the velocities of objects orbiting the galaxy, far from its center, should fall off with distance according to Kepler's law. As an example, a distant planet like Pluto (the largest and 2nd most massive dwarf planet in the solar system and the 9th largest and 10th most massive object directly orbiting the sun) moves at only 2.90 miles per second, whereas the earth (the third planet) orbits the Sun at 18.64 miles per second. Surprisingly, it turns out that objects orbiting the galaxy at larger distances from galactic center move around more or less at the same velocity as objects much closer to the center.

This can only be accounted for, if the mass progressively increases with radius as we move out further away from the central region. But this matter does not radiate as most of the light is from the central region. So the conclusion is that 90% of the galaxy is DM. This also seems to be universally true for all types of galaxies. Earlier it was thought that DM may be very faint, but sensitive measurements across all wavelengths suggest that such objects can hardly account for 5% of DM.

It is now felt that DM is some new kind of particle which does not couple with radiation. Many people for some time suspected that a familiar particle such as the neutrino (resulting from beta-decay and nuclear reactions) known to have been abundantly present in the early phase of the universe could have clustered around galaxies and galaxy clusters and is probably the DM. But the deduced mass of neutrinos is so insignificant that these may account for less than 1% of DM. Consequently many new particle candidates, such as WIMPS (weakly interacting massive particles), axions, gravitinos, gluinos, Q-balls, glue balls, and wimpzillas, were proposed in the quest of an explanation of the presence of DM—the most dominant mass of the universe.

These particles should have been produced in extreme energy conditions in the early universe and should now hang around as DM. The main purpose of the LHC (which mimics the energy when the universe was 1 ps old) is expected to produce such particles.

Further such particles are supposed to be ubiquitously present as DM, these could be detected in clever laboratory experiments. They should brush shoulders with normal atoms and nuclei generating detectable signals. Their momentum can be imparted by them to a nucleus or electron causing measurable small recoil. Consequently, several sophisticated experiments such as the DAMA, the Iodine, the Xenon, and the Cd-Te experiments are going on. These experiments are all underground so as to eliminate the effects of other particles like cosmic rays. All these experiments have not established so far the nature of the particle (or particles) making up DM. Several scientists meanwhile have pointed out that unusual gamma rays of energy of 50–60 GeV emanating from the galactic center could be evidence of annihilation of DM particles with their antiparticles. Their annihilation rate could be much higher since they are expected to be concentrated/clustered at the galactic center. This could also be explained as radiation coming, for instance, from high energy processes from pulsars. But the net outcome so far is that we are still ignorant about what DM really is.

5.27.4 Concept of zero and nonzero in western and Indian cultures

Western cultures have obviously had a concept of the void since antiquity. To express it, the Greeks had the word ouden ("void"). As for the Romans, they used the term vacuus ("empty"), vacare ("to be empty"), and vacuitas ("emptiness"); they also had the words absens, absentia, and even nihil (nothing), nullus and nullitas. But these words actually corresponded to notions that were understood very distinctly from each other. With the help of some appropriate examples, an etymological approach will enable us hereafter to form quite a clear idea of the evolution of the concepts down the ages and to perceive better the essential difference which exists between these diverse notions and the Indian concept of the zero.

"Presence" (from the Latin praesens, present participle of praesse, "to be before [prae]," "to be facing") is, properly speaking, the fact of being where one is. But the adjective present also means "what is there in the place of which one is speaking"; this meaning is applicable then both to an object and to a living being. In the figurative sense, applied to people, present means "that which is present in thought to the thing being spoken of" (to be present in thought at a ceremony, despite the physical absence); applied to things, however, it means "that which is there for the speaker, or for what he is aware of." It is thus a moral or deliberate presence.

Another meaning of presence, in opposition this time to the past and the future, is "that which exists or is really happening, either at the moment of speaking or at the moment of which one is speaking." Consequently, this meaning corresponds to the present situation. Figuratively, it is rather a matter of "that which exists for the consciousness at the moment one is speaking," somewhat like a scene one witnessed and which remains present on one's mind.

This preamble allows a better understanding of "absence," since it is a term that is opposed to presence. The word comes from the Latin absentia, which derives from abest, "is far." Thus it expresses the character of "that which is far from." It is thus by definition the fact of not being present at a place where one is normally is expected. And the absent is the person or the thing which is lacking or missing. As for nonpresence, it is simply the void left by an absence, since it is the space that is not occupied by any being or any thing. If it is an unoccupied place, it is this that is empty, whether it be a seat, an administrative post or even one of the "places" of the place-value system.

By dint of thinking solely of the void, some thinkers have arrived at vacuism, a type of physics, according to which there exist spaces where all material reality is void of all existence. It was developed notably by the Epicureans, who conceded the existence of places where all matter, visible or invisible, was absent. Others opted rather for antivacuism, like Descartes, who considered an absolute void to be a contradictory notion. It may be observed that "Torricellian vacuum" is not the absolute (mathematical) vacuum; this may be considered at best a numerical vacuum (just like numerical zero as opposed to exact zero).

Zero (Indian concepts of). In Sanskrit, the principal term for zero is Shunya, which literally means "void" or "empty." But this word, which was certainly not invented for this particular circumstance, existed long before the discovery of the place-value system. For, in its meaning as "void," it constituted, from ancient times, the central element of a veritable mystical and religious philosophy, elevated into a way of thinking and existing. The fundamental concept in *shunyatavada*, or philosophy of "vacuity," shunyata, this doctrine is in fact that of the "Middle Way" (Madhymakha), which teaches in particular that every made thing (samskrita), is void (shunya), impermanent (anitya), impersonal (anatman), painful (dukh), and without original nature. Thus this vision, which does not distinguish between the reality and nonreality of things, reduces the same things to total insubstantially.

This is how the philosophical notions of "vacuity," nihilism, nullity, nonbeing, insignificance, and absence, were conceived early in India (probably from the beginning of the Common Era), following a perfect homogeneity, contrary therefore to the Graeco–Latin peoples (and more generally to all people of Antiquity) who understood them in a disconnected and totally heterogeneous manner.

The concept of this philosophy has been pushed to such an extreme that it has been possible to distinguish twenty-five types of shunya, expressing thus different nuances, among which figure the void of nonexistence, of nonbeing, of the unformed, of the unborn, of the nonproduct, of the uncreated or the nonpresent, the void of the nonsubstance, of the unthought, of immateriality or insubstantiality; the void of nonvalue, of the absent, of the insignificant, of little value, of no value, or nothing, etc. In brief, the zero could have hardly germinated in a more fertile ground than the Indian mind.

Once the place-value system was born, the shunya, as a symbol for the void and its various synonyms (such as absence and nothing), naturally came to mark the absence of units. It is important to remember that the Indian place-value system was born out of a simplification of the Sanskrit place-value system as a consequence of the suppression of the word–symbols for the various powers of 10. This was a decimal positional numeration which used the nine ordinary names of numbers and the term shunya ("void") as the word that performs the role of zero.

Thus the Indian zero has meant from an early time not only the void or absence, but also heaven, space, the firmament, the canopy of heaven, the atmosphere and ether, as well as the terms such as nothing, a negligible quantity, insignificant elements, the number nil, nullity, and nothingness. This means that the Indian concept of zero by far surpassed the heterogeneous notions of vacuity, nihilism, nullity, insignificance, absence, and nonbeing of all the contemporary philosophies.

The Sanskrit language, which is an incomparably rich literary instrument, possessed more than just one word to express "void." It possesses a whole panoply of words which have more or less the same meaning; these words are related, in a direct or indirect manner, to the universe of symbolism of Indian civilization.

Thus words which literally meant the sky, space, the firmament or the canopy of heaven came to mean not only the void but also zero. In Indian thought, space is considered as the void which excludes all mixing with material things, and, as an immutable and eternal element, is impossible to describe. Because of the elusive character and the very different nature of this concept as regards ordinary numbers and numerals, the association of ideas with zero was immediate.

Other Indian numerical symbols used to mean zero were the terms such as purna (fullness, totality, wholeness, and completion), jaladharapatha (voyage on the water), and vishnupada (foot of Vishnu). To find out more about this symbolism, see the appropriate entries. Such a numerical symbolism has played a role that has been all the more important in the history of the place-value system because it is in fact at the very origin of a representation that we are very familiar with.

The ideas of heaven, space, atmosphere, firmament, etc., have been used to express symbolically, as has just been seen, the concept of zero itself. And as the canopy of heaven is represented by human beings either by a semi-circle or a circular diagram or again by a complete circle, the little circle that we know has thus come, through simple transposition or association of ideas, to symbolize graphically, for the Indians, the idea of zero itself.

It has always been true that "The circle is universally regarded as the very face of heaven and the Milky Way, whose activity and cyclical movements it indicates symbolically." And so it is that the little circle was put beside the nine basic numerals in the place-value system, to indicate the absence of units in a given order, thereafter acquiring its present function as arithmetical operator (that is to say that if it is added to the end of a numerical representation, the value is thus multiplied by 10).

The other Sanskrit term for zero is the word bindu, which literally means "dot." The dot, it is true, is the most elementary geometrical figure there is, constituting a circle reduced to its center. For the Hindus, however, the bindu (in its supreme form of a paramabindu) symbolizes the universe in its nonmanifest form and consequently constitutes a representation of the universe before its transformation into the world of appearances (rupadhatu). According to Indian philosophies, this uncreated universe is endowed with a creative energy capable of engendering everything; it is thus in other terms the causal point whose nature is consequently identical to that of "vacuity."

Thus this natural association of ideas with this geometrical figure, which is the most basic of them all, yet capable of engendering all possible lines and shapes (rupa). It is the perfect symbol for zero, the most negligible quantity there is, yet also and above all the most basic concept of all abstract mathematics. Thus the dot came to be a representation of zero (particularly in the Sharada system of Kashmir and in the notations of Southeast Asia), which possesses the same properties as the first symbol, the little circle.

This is the origin of the eastern Arabic zero in the form of a dot: when the Arabs acquired the Indian place-value system, they evidently acquired zero at the same time. This is why, in Arabic writings, sometimes the sign is given in the form of a dot, sometimes in the form of a small circle. It is the little circle that prevailed in the West, after the Arabs of the Maghreb transmitted it themselves to the Europeans after the beginning of the twelfth century.

To return to India, this notion was gradually enriched to engender a highly abstract mathematical concept, which was perfected in Brahmagupta's time (30 BC), that of the "number zero" or "zero quantity." It is thus that the shunya was classified henceforth in the category of the samkhya, that is to say the "numbers," so making the birth of modern algebra. So, from abstract zero to infinity was a single step which Indian scholars took early and nimbly.

The most surprising thing is that amongst the Sanskrit words used to express zero, there is the term Ananta, which literally means "Infinity." Ananta, according to Indian mythologies and cosmologies, is in fact the immense serpent upon which the god Vishnu is said to rest between two creations; it represents infinity, eternity, and the immensity of space all at once. Sky, space, the atmosphere, the canopy of heaven were, it is true, symbols for zero, and it is impossible not to draw a comparison in these conditions, between the void of the spaces of the cosmos with the multitude represented by the stars of the firmament, the immensity of space and the eternity of the celestial elements. As for the ether (dkasha), this is said to be made up of an infinite number of atoms (anu, paramanu). This is why, from a mythological, cosmological, and metaphysical point of view, the *zero and infinity have come to be united, for the Indians, in both time and space*.

But from a mathematical point of view, however, these two concepts have been very clearly distinguished. Indian scholars having known that one equaled the inverse of the other.

To sum up, the Indians, well before and much better than all other peoples, were able to unify the philosophical notations of void, vacuity, nothing, absence, nothingness, nullity, etc. They started by regrouping them (from the beginning of the Common Era) under the single heading shunyata (vacuity), then (from at least the fifth century CE) under that of the shunyakha (the sign zero as empty space left by the absence of units in a given order in the place-value system) before recategorizing them (well before the start of the seventh century CE) under the heading of shunya–sankhya (the "zero" number). Once again this indicates the great conceptual advance and the extraordinary powers of abstraction of the scholars and thinkers of Indian civilization.

The contribution of the Indian scholars is not limited to the domain of arithmetic; by opening the way to the generalizing idea of number, they enabled the rapid development of algebra and consequently played an essential role in the development of mathematics and all the exact sciences. It is impossible to exaggerate the significance of the Indian discovery of zero. It constituted a natural extension of the notion of vacuity, and gave the means of filling in the space left by the absence of an order of units. It provided not only a word or a sign, it also and above all became a numeral and numerical element, a mathematical operator and a whole number in its own right, all at the point of convergence of all numbers, whole or not, fractional or irrational, positive or negative, algebraic or transcendental.

5.27.5 Fast computation: Vedic way, by living and digital computers, and numerical zero consciousness

Computation is any type of calculation/process including nonnumerical one. The term "fast computation" implies, to a common human being, a computation that is carried out quickly (i.e., in a short time) and not slowly. This meaning is primarily true with respect to both living and nonliving computers.

As a technical term, a "fast computation" in the realm of computer science and for that matter in any science/engineering implies a computation which is carried out in time that is a polynomial (i.e., a finite degree polynomial in contrast to an infinite degree polynomial or an exponential function) of the problem size n. In contrast, a slow computation implies a computation that is carried out in time which is an exponential function (a polynomial of degree infinity) of the problem size n.

In fact, we count all the arithmetic operations (add, subtract, multiply, and divide operations), all control operations, such as testing whether a number is greater than another number, and jumping/branching, all input and output operations, and other associated system operations/commands. All these put together constitute computational complexity in a digital computer environment. Hence computational complexity and time complexity (time required to execute an algorithm) are directly proportional.

We may compare two or more algorithms in terms of time complexity which can be accurately and readily obtained in many software programming languages such the Matlab programming language. However, the most dominant term in the k-th degree polynomial is n^k, where n (considered large since for small n, the computational or the time complexity is practically a trivial issue) is the size of the problem.

For instance, the matrix multiplication of two square matrices, each of order n, when it is done by the conventional row-column multiplications, is considered a fast computation since it takes n^3 multiplications, that is, the computational time is proportional to n^3, that is, the time = f(n) which is expressed as $O(n^3)$ that has a form $an^3 + bn^2 + cn + d$ viz. a polynomial of degree 3, where O denotes 'of the order of' and a, b, c, d denote constants whose values depend on the specific algorithm used and the concerned computer. Since n^3 is the most dominant term, that is, n^3 is much larger than n^2 and n (as n is considered large, for small n, a fast and a slow algorithm make no significant difference), we neglect both n^2 and n and express the computational complexity or, equivalently, the cost of computation as $O(n^3)$.

If $n = 1000$, then both $n = 1000$ and $n^2 = 10^6$ are neglected when compared to $n^3 = 10^9$. Even in fast computation for the same problem, some algorithms are faster than others. A slow computation, on the other hand, is due to the use of a slow algorithm, that is, an algorithm which is exponential or, equivalently, combinatorial.

Consider, for instance, the traveling salesman problem. The problem states: Let there be n cities. A salesman starts from a city, visits each of the remaining $n - 1$ cities only once and then returns by the shortest route to the city where he started from. To solve this problem using exhaustive search needs the evaluation of 99! path computation when $n = 100$. The exhaustive search algorithm is combinatorial and hence exponential since by the Stirling's approximation $n! \approx \frac{\sqrt{2\pi n}}{1}(n/e)^n$, where $(n/e)^n = e^{n\log(n/e)}$. We have

99! ≈
9332621544394409609116046877265294032376216541518263059393616300788621604245200038032851664809578610057013175106171872672176235781392629880573437840656957744 ≈ 9.3326e + 155

The paths, which have 156 digits and which has been computed by using Matlab variable precision arithmetic command vpa(factorial(99), 157) or, equivalently, vpa(factorial(99), 200). Computation of these paths is simply intractable even in the fastest available computer (over 10^{18} flops, i.e., floating operations per second) today (2015). It can be seen that the 2014 Matlab can compute factorial of maximum 21 exactly (else it will be approximate). If computation of one path takes 1 μs, then the computation of 99! paths would take 99! microseconds, which is .29593548783594652489586652959365e143 years or approximately 3×10^{142} years—an astronomical figure (beyond imagination)!

If, on the other hand, $n = 5$ (i.e., only 5 cities) then we need to evaluate only 24 paths, which is immediate (within a second) for the slow exhaustive (deterministic) search algorithm. There exists so far no polynomial-time algorithm for the traveling salesman problem, nor has anybody so far proved that no polynomial algorithm could exist for the traveling salesman problem. Neither does there exist a polynomial-time algorithm to verify a given answer as to whether it truly corresponds to the minimal cost path. If you can innovate one deterministically in polynomial-time, then you will definitely raise the humanity to a much higher level. Such an innovation is much more important than most of the Nobel laureates' contributions (put together) so far in the history. You would save trillions of dollars per year to the world!

The computational complexity in the realm of living computers is not precisely understood/defined. Even a living computer, for example, a super-human being with extraordinary computing capability may not be able to come out and tell others how he performs the computations involving many arithmetic operations in an extraordinarily short time. The whole computational aspect in the mental plane will continue to remain unclear and possibly nonunique, that is, the aspect varies from person to person.

The numerical computational procedure embodied in vedic mathematics happen to be not only of the precomputer era but also of practically the prehistoric era developed in India. Vedic computation is meant for bigger numbers—larger than we study in a multiplication table in school, but smaller than practiced by a few human computers in the world. This computation practiced by a human has the limitation that like any computer it cannot usually take very large numbers and many of them. The complexity of computation implementing vedic computational procedures on a digital computer is usually larger than what a conventional computational procedure or an intelligent computational procedure on a digital computer has.

If you ask a student to find the square-root of a 3- or 4-digit positive integer, he will simply take out his calculator and tell you the value calculated by the calculator without much of a feeling for the magnitude. Everything seems to happen in a quite mechanical way in which consciousness is low or a numerical zero (not exactly zero since exact zero consciousness is nonexistent). A vedic arithmetician on the other hand consciously would apply the vedic computational rules and obtain the result. As a result he is not usually oblivion about the magnitude of the quantity that he gets and check if the result is reasonable. Thus major error may be averted in this by the people practicing vedic way of computation. A vedic arithmetician usually will be a common human being with no super-human computing power unlike human computers such as Johann Martin Zacharias Dase and Shakuntala Devi (discussed later).

5.27.6 Vedic mathematics versus computer mathematics

With regard to a living computer (e.g., living human being) and nonliving computer (e.g., digital computer), each of the two mathematics has its own scope. In the realm of vedic mathematics (specifically arithmetic), a living computer does the computation in the mental plane and is faster than in the conventional mathematics. Vedic mathematics comprises a number of vedic algorithms which can be implemented on a computer, a very general algorithm involving arithmetic with truly large numbers, say 20 digit by 25 digit needs to be devised/written in vedic mathematics as such an algorithm may not be readily available.

Some recognition problems are solved much faster by a living computer than by a modern digital computer. Such a recognition problem is usually not in the realm of vedic mathematics. So far as the computational complexity is concerned, computation with vedic mathematical methods on a digital computer has higher time complexity than that with conventional mathematical methods. That is one of the reasons why vedic mathematical methods have not become popular on a digital computer.

On the other hand vedic mathematical methods on a living computer has a lower complexity than the corresponding complexity with conventional mathematical methods. Unlike the algorithms executed by Shakuntala Devi, which is unknown to a scientist and even perhaps to Shakuntala Devi, the vedic algorithms are explicitly known by a scientist. In some cases (e.g., the case of Shakuntala Devi) this computational (time) complexity in the mental plane is lower than that in the case of a digital computer, but in other cases, this could be practically infinitely faster than the fastest digital computer.

It may be seen that the natural mathematics works on real numbers in parallel while both vedic mathematics and conventional mathematics work on rational numbers with always a finite precision sequentially or partially parallel (on a digital computer). One aspect of using vedic mathematics is that it probably injects one with a better idea/appreciation of the magnitude of the result. Consequently, one is not going to produce a wild result far from the true result/solution. So far as the mistakes are concerned, a living computer without any exception is prone to committing mistake(s) while a digital computer is not prone to committing mistake(s).

5.28 Storage capacity and computational power of nonliving and living computers

We put forth the way a computer scientist/a neuro-physicist look at the problem of a living computer, for example, a living human being. According to them, neurons = (nerve) cells that transmit information/data in a living being's nervous system so that it can detect stimuli from its environment and act accordingly. Not all living beings have neurons. Sponges and *Trichoplax*, for example, do not have neurons. Neurons may be packed to constitute a structure such as the brain of vertebrates (= animals with internal skeleton made of bone) and neural ganglions (nerve cell clusters located in the peripheral nervous system) of insects. A nonliving computer primarily implies here a digital computer although. an analog computer (= the computer that measures numerically continuous physical quantities such as temperature, pressure, length, electrical current, and voltage) is also nonliving and when it is considered, it is explicitly mentioned as "analog computer." Table 5.1 provides the storage capacity and computing power of living and nonliving computers.

5.28.1 Limit of computation by a nonliving computer

Every 18 months processor speed is doubling. Every 12 months bandwidth is doubling and every 9 months hard disk space is doubling. Behind all this exponential growth is the computational mathematics superimposed on the hardware. In this context, the following story is significant. A merchant asked his friend (another merchant) to supply matchsticks for 1 month—1 matchstick on Day 1, 2 on Day 2, 4 on Day 3, 8 on Day 4 and so on for 30 days. For this 1 month's supply, he would pay his friend Rs. 100,000. The friend was very glad to readily accept the order as he felt it was indeed a very profitable offer without realizing that the supply grows exponentially and consequently it soon turns out to be beyond his means. In fact, the friend needs to supply

$$2^0 + 2^1 + 2^2 + 2^3 + 2^4 + \cdots + 2^{28} + 2^{29} = 1 + 2 + 4 + 8 + 16 + \cdots + 268435456 + 536870912 = 1073741823$$

matchsticks, computed using the Matlab commands *> > s = 0; for i = 0:29, s = s+ vpa(2^i), end;*

If a matchbox consisting of 50 sticks costs Rs. 5, then 1073741823 sticks would cost Rs. 107374182.3 which is 1073.741823≈1073 times larger than what the merchant had offered!

The question now arises: Is there any limit/barrier to increase the computational power arbitrarily?

Table 5.1 **Storage and computational power of living and nonliving computers**

Computer (Living/nonliving)	Neurons in brain/ whole nervous system	Neurons in cerebral cortex	Storage capacity (no. of bits)	Computational power (no. of bits/s)	Synapses[a]
Abacus	0	0	10^0	10^0	10^0
Radio channel	0	0	10^0	10^3	10^0
Television channel	0	0	10^0	10^6	10^0
Sponge[b]	0	0	10^0	10^0	10^0
Trichoplax[c]	0	0	10^0	10^0	10^0
Viral DNA			10^3	10^0	
Hand calculator	0	0	10^3	10^3	10^0
Smart missile[d]	0	0	10^3	10^9	10^0
Bacterial DNA			10^6	10^0	
Bacterial reproduction			10^6	10^3	
Personal computer	0	0	10^6	10^6	
Mainframe computer (1980s)	0	0	10^8	10^8	
Human DNA			10^9	10^0	
Roundwarm	302				
Jelly fish	800				
Medicinal leech	10,000				
Pond snail	11,000				
Sea slug	18,000				
Fruit fly	100,000				
Lobster	100,000				
Ant	250,000				
Honey bee	960,000		10^9	10^8	
Cockroach	1,000,000				

Frog	16,000,000				
Mouse	71,000,000		10^{9}	10^{10}	
Rat	200,000,000		10^{9}	10^{10}	
Octopus	300,000,000				
Telephone system	0	0	10^{11}	10^{13}	
English Dictionary	0	0	10^{12}	10^{0}	
Video recorder	0	0	10^{12}	10^{6}	
Cray supercomputer (1985)	0	0	10^{12}	10^{11}	
Human visual system			10^{13}	10^{13}	
Supercomputer (2003)	0	0	10^{14}	10^{13}	
Elephant			10^{14}	10^{16}	
Human being	86,000,000,000–114,000,000,000		10^{14}	10^{13}	10^{14}–10^{16} (av. adult)
British Library	0	0	10^{15}	10^{0}	
Whale			10^{16}	10^{16}	

[a]Synapse= structure that permits a neuron pass an electrical or chemical signal to cell (neural or otherwise).
[b]Sponge= animals with multicellular organisms that have bodies full of pores and channels allowing water to circulate through them.
[c]*Trichoplax*=animals with a basal group of multicellular organisms that are very flat having a diameter of around 1 mm.
[d]Smart missile= guided missile intended to precisely hit a specific target with minimum or zero collateral damage.

The answer is "yes." It is physics that sets up the limits.

i. *Speed of light barrier* Electrical signals (pulses) cannot propagate faster than the speed of light. A random access memory used to 10^9 cycles/s (1 GHtz) will deliver information at 0.1 ns (0.1×10^{-9} s) speed if it has a diameter of 3 cm.

ii. *Thermal efficiency barrier* The entropy of the system increases whenever there is information processing. Hence the amount of heat that is absorbed is kT $\log_e 2$ per bit, where k = Boltzmann constant = 1.38×10^{-16} erg per degree and T is the absolute temperature (taken as room temperature) = 300 K. It is not possible to economize any further on this. If we want to process 10^{30} bits/s, the amount of power that we need is p = $10^{30} \times k \times 300 \times \log_e 2$ = 28696293275181728 erg/s obtained using the Matlab variable precision arithmetic (vpa) command

```
>> p = vpa(10^20*1.38*10^ - 16*300*log(2), 40)
```

This amounts to p = 2869629327.5181728 W≈2.87×10^9 W, where 1 W = 10^7 erg/s.

iii. *Quantum barrier* Associated with every moving particle is a wave which is quantified such that the energy of one quantum E = hν, where h = Plank's constant and ν = frequency of the wave. The maximum frequency $\nu_{max} = mc^2/h$, where m = mass of the system and c = velocity of the light. Hence the frequency band that can be used for signaling is limited to the maximum frequency ν_{max}. From Shannon's information theory, the rate of information (number of information that can be processed per sec) cannot exceed ν_{max}. The mass of hydrogen atom is 1.67×10^{-24} gm, c = 3×10^{10} cm/s, h = 6×10^{-27}. Thus per mass of hydrogen atom, maximum $1.67 \times 10^{-24} \times 3^2 \times 10^{20}/(6 \times 10^{-27}) = 2.5050 \times 10^{23}$ bits/s can be transmitted.

The no. of atoms in the observable universe is estimated to be around 10^{80}. Thus if the whole universe is dedicated to information processing, that is, if all the atoms are employed to information processing simultaneously (parallely), then no more than 2.5050×10^{103} bits/s or 2.5050*10^103*60*60*24*365 = 7.8998e + 110 = 7.8998×10^{110} bits/year can be processed. This is the theoretical limit (for massively parallel processing) set by the laws of physics and we are nowhere near it!

A single processor (atom) can process only 2.5×10^{23} bits/s or 7.9×10^{30} bits/year and no more!! (Question: Assume that every 18 months, a processor speed is doubling. If the CPU clock speed (i.e., clock rate) is 8.31 GHz and if we assume that the bit rate (bits/s) = clock rate, then what will be the processing speed of the CPU? How many years will be required to reach the theoretical limit?)

5.28.2 Limit of computation by a living computer

The term "living computer" implies any living being such as a living human being, a living bird, a living animal, a living insect, and a living bacteria. A living computer, for instance, a living human being, has 100 billion neurons (on average). If it is assumed that a neuron can hold one bit (bit=binary digit 0 or 1 constitutes the building block of any information, i.e., a bit cannot be subdivided) of information, then a man will have 10^{11} bits of information stored in his memory, which is fairly large compared to information stored in a public library.

If a neuron has to hold information then it has to be at least one bit of information since anything less than one bit is simply zero bit (i.e., no information). That is the best a computer scientist could assume. It is the fact that a human being has a finite number of neurons. But it is definitely not the fact that the information content and hence the knowledge content in the mind of a living human being is finite. As a matter of fact the information/knowledge that exist in one's mind is limitless (infinite). There exists no knowledge outside one's own mind.

This implies that the knowledge existing in Mr X's mind is exactly equal to that existing in Mr Y's mind. The only difference is the speed of mining a specific knowledge from one's own mind. A scientist (materials/spiritual) can only do knowledge mining in the ocean of knowledge that is already there in his/her mind. In the absolute sense nobody creates any knowledge not already existing in him. The proof is not mathematical in which fuzziness cannot be ruled out. The proof is through actual deep meditative state of the mind. Our rishis (spiritual scientists of the world) have experienced it.

This first-hand experience is the best possible proof (better than any mathematical proof based on deduction/induction/contradiction/construction or any other method) totally devoid of fuzziness. The biological neural network responsible for communication has a speed based on a state of mind out of possible infinite states of mind, which can be categorized in four major divisions, that is, superconscious, conscious, subconscious, and unconscious (consciousness not exactly zero). In each of these categories, there are infinity of states.

The speed of neural communication is extremely fast, possibly faster than the *fastest* available digital computer today (2015), in the unconscious state and not so prone to mistakes/errors while it is much slower in the conscious state of mind—much slower than an ordinary laptop and considerably prone to mistakes/errors. The speed and quality of transmission in the subconscious state will be somewhere in between those of conscious and unconscious states. The probability of committing an error in any state of mind of any living being is never zero (in this regard there is no exception).

This probability depends on a state of mind and varies from one state to another while the probability of committing an error (here "error" implies "mistake") by a nonliving modern computer is zero. Thus we have the proverb: *To err is human. Not to err is computer.*

5.28.3 Extraordinary living computers

All living beings including animals and birds do compute consciously, unconsciously, or subconsciously, a few of them have extraordinary computing capability. We give below a couple of examples of living human computers who, unlike a nonliving modern computer, are not completely mistake-free (no living being can be always error/mistake-free).

Johann Martin Zacharias Dase (1824–1861) was born in Hamburg, Germany. He was a calculating prodigy. (Prodigies have extraordinary ability in some area of mental calculation, such as multiplying large numbers, factoring large numbers, or finding roots of large numbers.) Johann attended schools in Hamburg, but he made only a little progress there. He used to spend a lot of time developing his calculating skills; people around Johann found him quite dull. He suffered from epilepsy throughout his life, beginning in his early childhood. At the age of 15 he gave exhibitions in Germany, Austria, and England.

His extraordinary calculating powers were timed by renowned mathematicians including Gauss. (Dase multiplied 79,532,853 by 93,758,479 in 54 seconds; two 20-digit numbers in 6 minutes; two 40-digit numbers in 40 minutes; and two 100-digit numbers in 8 hours 45 minutes.) In 1840, he struck up an acquaintance with L.K. Schulz von Strasznicky (1803–1852), who suggested that he apply his powers to scientific purposes. When he was 20, Strasznicky taught him the use of the formula

$$\pi/4 = \tan^{-1}(1/2) = \tan^{-1}(1/5) + \tan^{-1}(1/8)$$

and asked him to calculate π. In 2 months he carried the approximation to 205 decimal places, of which 200 are correct. He then calculated a 7-digit logarithm table of the first 1,005,000 numbers during his off–time from 1844 to 1847, while he was also occupied by work on the Prussian survey.

His next contribution was the compilation of a hyperbolic table in his spare time, which was published by the Austrian Government in 1857. Next, he offered to make a table of integer factors of all numbers from 7,000,000 to 10,000,000; on Gauss' recommendation, the Hamburg Academy of Sciences agreed to compensate him, but Dase died shortly thereafter in Hamburg. He had an uncanny sense of quantity; he could just tell, without counting, how many sheep were in a field or words in a sentence, and so forth, up to quantities of about 30.

Dase may not be aware about how he does the computations in his mind. Possibly besides a conscious state of mind he would be using other nonconscious state also. Several other calculating prodigies have been discovered, such as Jedediah Buxton (1707–1772), Thomas Fuller (1710–1790), Richard Whately (1787–1863), Zerah Colburn (1804–1840), George Parker Bidder (Andrè–Marie Ampère, 1806–1878), Truman Henry Safford (1836–1901), Jacques Inaudi (1867–1950), and Shakuntala Devi (1929–2013) of India (who appears to be faster than the fastest computer on earth).

The case of *Shakuntala Devi* is different from that of Dase. On one day in 1970s, she gave a talk and demonstrated her computational prowess in the Lecture Hall No. 303 of the Power Engineering Building of Indian Institute of Science (IISc), Bangalore. Earlier she had shown her capability of computing the n-th root of a large positive integer and also of multiplying two large numbers in several universities in Europe. All these appeared to her to be "child's play."

However, a graduate student of computer science and automation of IISc wrote a huge arithmetic expression involving terms such as $52.345^{-4.71}/26.81^{7.2}$ and running from top left corner to the bottom right corner of the fairly long black board. The student had already computed the value of the expression using the then IBM 360/44 (mainframe) computer in IISc. Shakuntala Devi read the expression sequentially while the student was writing and then within about 2 s she told the answer to the audience. The student responded saying that the answer was wrong. She was startled for a moment and almost immediately came out with the answer which was amazingly correct!

Nobody (perhaps including she) knows how she could do the computation so fast. Probably other states (other than her conscious state) of her mind are involved in the computation. So far as Dase is concerned, he would possibly be using more of his conscious state (than Devi's) in addition to other states. This is one of the author's (Sen's) first-hand experiences. She claimed that she is faster than the fastest computer in the world! This has not been disputed anywhere although her speed and the concerned computer's speed have not been compared formally by us. She is called a human computer. Definitely she has a good knowledge of arithmetic computation. While the proverb viz. *To err is human* is eternally true, *Not to err is computer* is also true, where "err" implies "mistake."

5.28.4 Which is faster—living computer or nonliving computer?

Around 1980 one of the authors (Sen) had a chance meeting with Prof. Satish Dhawan inside the State Bank of India, Indian Institute of Science (IISc) campus, Bangalore. He was then the Director of IISc (and also the Chairman of Indian Space Research Organization (ISRO)) and Sen was a faculty member of the then Computer Centre (now called Supercomputer Education and Research Centre) of IISc. Sen was then designing and developing computer programs for research problems of various engineering departments as well as of the then department of Applied Mathematics on a mainframe computer in IISc and also that in Tata Institute of Fundamental Research, Mumbai. Many PhD research students considered him a good programmer and used to come to him to take his computational help.

Prof. Dhawan asked him: Tell me, Sen, which is faster—a supercomputer or a human being in computation? He responded readily saying that definitely "a supercomputer" and felt happy believing that he could answer correctly the too easy a question. By the time one multiplies manually a 3 digit by another 3 digit number, the result of which may not be always correct, the computer will compute around one hundred thousand multiplications each of 15 digit by 15 digit without any mistake. He did not realize then that he had given an answer rather too rashly. Prof. Dhawan explained: No, Sen, a human being is much faster than a supercomputer. Take, for instance, the photograph of your mother. How much time do you take to respond that it is your mother's photograph? A fraction of a second. But when you use Cray supercomputer, it would take several seconds to come out with the correct answer.

Sen's implicit ego that he was good in programming was shattered. He started thinking deeply for several days to find out the reason why he committed such a blunder. It appeared to him that there are various (numerous, rather infinite) states of mind—broadly super-conscious, conscious, subconscious, and unconscious states (each one of these major four states has infinity of states). In conscious state we are slow (think mostly sequentially with full consciousness) and more prone to error while in subconscious state, we are faster (think considerably parallely with some consciousness) and less prone to error and in unconscious state, we are fastest (think mostly parallely without being much conscious of our action) and least prone to error. In super-conscious state, we are superfast and almost error-free.

As a driver of a car, when we apply break, we change the gear without being conscious of it. If we attempt to do these consciously then we will be slow and also more likely to commit error (such as pressing the accelerator rather than the brake in a hurry). When one sees his mother's photo, many parallel actions take place very fast, which he is not aware about nor is he knowledgeable about how the whole process of recognition proceeds.

Most of the revelations take place in the super-conscious state of the mind. A scientist is an intense thinker and when he focuses his mind to a specific question/problem, he often gets the answer/solution. In fact all the knowledge exists in one's mind. All he does is *knowledge mining*.

So far as the speed is concerned, at some state of mind, the speed of a living computer is theoretically infinite. There is yet no way to formally compute this speed since we do not know how the biological neural communications at any state of mind occur. In some state of mind, the communications are relatively slow, even much slower than a laptop while in some other state it is much faster, even faster than the fastest available computer (over 10^{18} flops) on earth today (2015) or in future.

5.28.5 Which has more storage—living computer or digital computer?

So far as the storage is concerned, a living computer such as a living human being has an infinite storage (although the biological neuron count is around 100 billion according to neuroscientists/physicists). We do not know how the information is stored in a neuron nor do we know the storage capacity of a neuron is truly infinite. Nevertheless an ocean of information resides in one's mind and it is observed/experienced in deep meditation to be limitless. One's mind is the store house of the complete ocean of knowledge. One's job is to retrieve a specific information/knowledge.

How fast one can access this depends on one's state of mind. This state varies from one moment to another. There exists no knowledge outside one's (anybody's) own mind. For a digital computer, the storage is finite and will continue to remain ever finite although it increases exponentially with time with an upper limit.

5.29 Ramanujan versus Devi and Dase

We believe that there are infinity of states of mind of a living computer, that is, a living human being. We may divide these states into four categories—unconscious (consciousness is not exactly zero but low), subconscious, conscious, and super-conscious. Each of these categories has (evidently) infinity of possible states as earlier stated. The majority of the common men display or implicitly use certain states in each of the four levels. When one uses states beyond these certain states, then he turns out to us as one distinctly different from a common human being in the context of extraordinary mental ability. These mental ability obviously makes use of the knowledge existing in some of the states of conscious, subconscious, and possibly super-conscious states.

Ramanujan (1887–1920) differs from Shakuntala Devi both in the speed of arithmetic computation and in the retrieval of background knowledge regarding a subject. While possibly Shakuntala Devi depicted extraordinary speed of arithmetic computation involving many numbers including fractional ones and their powers/fractional powers and excelled Ramanujan (and perhaps other few human computers) in this regard, Ramanujan has definitely excelled Shakuntala Devi in numerical/mathematical knowledge and its interpretation (including spiritual ones) that is truly amazing.

The notebooks/scribbling of Ramanujan traced and being researched has been revealing the astonishing mental ability of Ramanujan, which is very much well above others concerned. People are still trying to understand Ramanujan and his very intimacy with numbers (as if these are living beings and carry important messages) and getting deeper insight into the mind of this genius. So far as Dase is concerned, he is significantly different from both Shakuntala Devi and Ramanujan. He used states of his mind, which are not identical to those of Devi and Ramanujan, but these states are definitely extraordinary. Further, numbers may not be his personal friends as are so with Ramanujan. The relationship among numbers and the highly interesting aspects of many of them seem to be an integral part of Ramanujan's mind, but these do not appear to be so with both Devi and Dase.

Other aspects of Ramanujan's life that seem to differ from other humans with extraordinary computing and other capabilities may be depicted through the well-known and well-researched activities described in the previous section.

5.29.1 Ramanujan: thought-provoking questions on 1729, 0, ∞, computation of Pi and Theta

Divinity was manifested through Srinivasa Ramanujan (1887–1920) in terms of numbers. We represent some of the numbers and their interesting and deeply significant spiritual interpretation by Ramanujan. In other words, the nonconventional way Ramanujan has viewed/experienced the numbers is simply amazing. One needs to ponder deeply on whatever has been revealed and uttered by him. To him, each number is a living being and his personal friend carrying important distinct messages. This is, unlike us, the common human beings, who do not feel/view the numbers as beings with interesting characters and significant message carriers and as those which evoke a personal relationship with our emotional attachments.

By computation, we imply any type of calculation/process including nonnumerical one.

1. Hardy–Ramanujan Number: $1729 = x^3 + y^3 = u^3 + v^3$. x = ? y = ? u = ? v = ?
   where x, y, u, and v are all 1- or 2-digit distinct integer. (*Answer without seeing internet or book.*) Use Matlab/any means to find/compute next pair of any k-digit integers (x, y), k >2.
2. 0 = God (Absolute Reality), ∞ = His (God's) manifestation. $0 \times \infty$ = not one Number but all Numbers, each of which corresponded to individual acts of creation. Imagine (seriously) everything including your own body is vanishing. Will it be dreadful to you? (Thou shalt not divide by zero. Natural Mathematics/Computation has no division by zero.)
3. Compute π, correct at least up to seven significant digits, using the identity due to Ramanujan

$$\pi = 1/[(2 \times \sqrt{2}/9801)\sum_{k=0}^{\infty}(4k)!(1103 + 26390k)/(k!^4 396^{4k})]$$

   taking, for ∞, (i) 5 and (ii) 6. Hence compute the relative error in π.

4. Ramanujan's general theta function f(a, b) is defined by

$$f(a,b) = \sum_{k=-\infty}^{\infty} a^{k(k+1)/2} b^{k(k-1)/2},$$

where |ab| <1. Compute, using Matlab/any means. (i) f(0.5, 0.4), (ii) f(0.9, 1), and (iii) f(0.01, 0.02) correct up to four significant digits. Take, for ∞, appropriate (minimum) number (of terms) so that you get the required accuracy. Compute the relative error in computing f(a, b).

5.29.2 Squares summation, ∞/non-∞, natural/artificial consciousness/intelligence, living/nonliving computers: questions

1. The sum of n + 1 (n > 0) consecutive squares starting with x = n(2n + 1) is equal to n consecutive squares starting with y = x + n + 1. For example, if n = 1, then x = 3, y = 5 and hence $3^2 + 4^2 = 5^2$. Take (i) n = 2, (ii) n = 3, and (iii) n = 4 and construct the identities.

2.

i. What is (are) the difference(s) between ∞ and non-∞? (∞)
ii. What is (are) the difference(s) between Natural Consciousness and Artificial Consciousness? (Natural consciousness is possessed by a living being while artificial consciousness is a simulation of natural consciousness and is depicted in a finite/limited way by a computer.)
iii. What is (are) the difference(s) between Natural Intelligence and Artificial Intelligence? (Natural intelligence is infinite/unlimited while artificial intelligence is not.)
iv. Which is faster—Living Computer (e.g., Living Human Being) or Nonliving Computer (e.g., Hyper (exa-flops)-computer)? (Living computer)
v. Is there any point in the infinite space/cosmos (universe), where consciousness is exactly zero? (No)
vi. What is consciousness? How do you visualize/distinguish consciousness of a living being and that of artifacts? Is the consciousness in a dead (decaying) human body exactly zero? (*Part 1.* There is no unique definition of consciousness. However, it may be viewed as the state of awareness/being aware of an external object or something within (oneself). *Part 2.* Consciousness of a living being is natural and much more vibrant than that of an artefact. *Part 3.* No, but very small. Perpetual/absolute inactivity is absent. Pure consciousness is omnipresent.).
vii. How many states of consciousness do we (living being) have? Is the number finite (say, 4) or infinite? Why?

5.29.3 Ramanujan versus Human Computers: question

How does Ramanujan differ from Human Computers such as Johann Martin Zacharias Dase (1824–1861) (who multiplied … two 40-digit numbers in 40 minutes; and two 100-digit numbers in 8 hours 45 minutes. He then calculated a 7-digit logarithm table of the first 1,005,000 numbers during his off–time from 1844 to 1847, …), Jedediah Buxton (1707–1772), Thomas Fuller (1710–1790), Andrè–Marie Ampère, Richard Whately (1787–1863), Zerah Colburn (1804–1840), George Parker Bidder (1806–1878), Truman Henry Safford (1836–1901), Jacques Inaudi (1867–1950),

Kanala Sriharsha Chakravarthy, and Shakuntala Devi (1929–2013) (who appears to be faster than the fastest digital computer on earth)?

5.30 Limitations of comprehension of a common human being

A common human being understands 7 ± 2 things at a time. If one reads out a telephone number 23372198, it will not be difficult for him to memorize and reproduce later. If, on the other hand, he/she reads out a 100-digit number to him, it will not be possible for him to remember and then reproduce later. For an uncommon/extraordinary human being, remembering a 100-digit (or bigger) number is possible.

One of the authors (Sen) personally came across a young high school Telugu boy, Kanala Sriharsha Chakravarthy, in December, 1994 at the 39th Congress of Indian Society of Theoretical and Applied Mechanics held at Andhra University (Waltair, Andhra Pradesh, India). A number more than 100 digits was read out to him and he then readily reproduced the number absolutely correctly. He told the audience that he could reproduce the number even after 1 week. He was capable of performing arithmetic computations too with a speed much beyond the capability of a common human being.

We have come across extraordinary human beings with different kinds of capabilities—not all are necessarily human computers or human beings with uncommon memories. There are personalities such as Albert Einstein and Srinivas Ramanujan who are capable of seeing the physical/mathematical world far beyond that of common humans. Each of these humans has excelled in one (and sometimes more) of the specific mathematical and computational sciences and earned for him/her a status as an extraordinary human or a genius. Many of them (not all) are able to bring down their findings/visions to a level which a sincere common man/seeker (of truth) can comprehend/understand. In other animal worlds, for example, in the world of whales/dogs, such capabilities do exist, but we are not enough knowledgeable about their uncommon animal behavior.

5.30.1 Infinity versus noninfinity

"Fast computation" invariably conjures up in a scientist (specifically a computer/computational scientist) a feeling of quick computation in contrast to slow computation. In this context, a digital computer is improving in its computational speed exponentially over certain periods of time, it remains ever finite though.

On the other hand, a living computer at certain state of mind depicts infinite computational speed, not scientifically explainable though. Shakuntala Devi may be considered as an example in this aspect. Thus the difference between speed of a living computer and a nonliving computer is infinite (i.e., infinity minus any finite quantity is always infinity). The difference between natural consciousness and natural intelligence (e.g., of a living human) and artificial consciousness and artificial intelligence (e.g., depicted by a computer) is infinite. The way Shakuntala Devi does the computation mentally seems to be different from that of Dase. *None of them has come out with the formal procedure or, equivalently, the algorithm that they execute in their mind. In all probability, their computational activities are in different mental states and none of them appears to come out with the exact/precise steps they follow in their mind.*

5.31 Consciousness of living beings versus that of artifacts

There is no precise single definition of consciousness. Like the definitions of mathematics, consciousness also is defined/viewed differently by different individuals based on their individual perception (of consciousness). Consciousness is defined as the state of awareness/being aware of an external object or something within (oneself). It is viewed as sentience, awareness, subjectivity, or to feel wakefulness having a sense of selfhood and the executive control system of the mind. Many scientists believe that experience is the essence of consciousness and it can only be completely known subjectively from the inside. If consciousness is subjective and not visible from the outside, why do the vast majority of people believe that other people are conscious, but stone and wood are not? This is known as the problem of other minds.

Consciousness has been considered out of bound of materials science such as physics and chemistry and has been avoided as an important research topic by scientists. This is because of a common feeling that a phenomenon defined in a subjective term cannot be meaningfully studied employing an objective experimental procedure.

A psychological study which differentiated between (i) fast, parallel, and extensive unconscious processes and (ii) slow, serial, and limited conscious processes was made by George Mandler in 1975. Since the 1980s, psychologists and neuroscientists developed an area of research called *consciousness studies* resulting many experimental and methodological work published in journals such as *Consciousness and Cognition* and *Journal of Consciousness Studies*. Moreover, regular conferences have been organized by societies such as the Association for the Scientific Study of Consciousness.

Consciousness of an artifact/a computer is just that consciousness that *Paramatma* (universal soul, i.e., the soul of the universe, that exists everywhere; there exists no point in infinite space where this consciousness is absent) has. On the other hand, *Jeevatma* (the individual soul) is having a consciousness much more vibrant than that of an artifact or a computer. However, the way we attribute consciousness, it is the consciousness that is exclusively possessed by a living being (e.g., a living human). And the consciousness depicted by an artifact is just a simulation of the natural consciousness; hence we can call it artificial consciousness which is no real consciousness of a living being.

Both artifacts and living beings have one thing in common, that is, enormous amount of activities (perpetual/absolute inactivity is absent in both) which are ever present to a varying degree. Vedic mathematicians are common living human beings and do computations using vedic rules usually consciously. They do not, however, have the ability of performing computations which are done by Shakuntala Devi. Their speed of computation cannot match that of Shakuntala Devi. Further, a vedic mathematician cannot compute a truly large arithmetic problem that can be readily solved by Devi. A difference is that a vedic mathematician consciously knows what he is doing step by step while Devi seems not to know the steps that she performs sequentially/in parallel.

Since consciousness is out of bound of materials science such as physics, it has limitations/constraints in terms of explaining the cause of nonstop activity that has been going on eternally throughout the infinite space/endless universe.

5.31.1 Pure consciousness: Eternal Witness

Pure consciousness (distinct from experienced (by mind) consciousness or a reflection of pure consciousness in the mind) has intrinsic existence, that is, it was existing everywhere infinite years ago, it is existing now, and it will never die. It remains constant, that is, it never changes. It is existence absolute. Our true nature or, equivalently our soul, is existence absolute, consciousness absolute, and bliss absolute. Our body does not have intrinsic existence (i.e., it is extrinsic). It comes and goes. Date of birth and date of death (of a person written on a tomb or a memorial) imply that our bodily existence is transient, but we are not. We do exist eternally (Sat-chit-ananda, the Sanskrit term, meaning "existence absolute, consciousness absolute, and bliss absolute"). Our soul, not that it exists, is existence itself. Our soul, not that it knows, is knowledge itself. We (our souls) always are our own continuous (permanent) source of the purest joy/the highest happiness. In other words, we (our souls) are the highest source of bliss.

A comprehensive technical manual *Panchadashi* (based on pure reasoning, not borrowing/citing the scriptures) written in the Sanskrit Sloka format over 700 years ago (English translation and notes by Swami Swahananda) by Sri Vidyaranya Swami, the spiritual head of Sringeri Math, Karnataka, India during 1377–1386 AD and published by Advaita Ashrama in 1967 is ideally suited for the modern man to understand "Sat-chit-ananda" of our true nature. However, to realize/experience this needs yogic practice/meditation for a common man (since logical understanding of our own divinity using reasons, i.e., *Apara vidya* is not enough; only experiencing the self is the true knowledge, i.e., *Para vidya* and is all that matters).

A Sufi (Sufism or Tasawwuf is defined as the inner mystical dimension of Islam) poet said: I look for Allah (God), I find myself. I look for myself and I find Allah, the Eternal Witness. How does pure consciousness relate to "zero"? It relates to zero in the same way as "infinity" relates to zero which is infinity's "twin," and the infinity is its (zero's) manifestation.

5.32 Why did the original name of zero come to be used for the whole set of Indo–Arabic numerals?

The answer lies in the attitude of the Catholic authorities to the counting systems borrowed from the Islamic world. The Church effectively issued a veto, for it did not favor a democratization of arithmetical calculation that would loosen its hold on education and thus weaken its power and influence; the corporation of accountants raised its own drawbridges against the "foreign" invasion; and in any case the Church preferred the abacists—who were most often clerics as well—to keep their monopoly on arithmetic. "Arabic" numerals and written calculation were thus for a long while almost underground activities. Algorists piled their skills in hiding, as if they were using a secret code.

All the same, written calculation (on sand or by pen and ink) spread amongst the people, who were keenly aware of the central role played by zero, then called cifra, or chifre, or chiffre, or tziphra, etc. By a very common form of linguistic development, known as synecdoche, the name of the part (in this case, zero) came to be used for the whole, as in a kind of shorthand, so that words derived from sifr came to mean the entire set of numerals or any one of them. Simultaneously, it also came to mean "a secret," or a secret code—a cipher.

So the history of words for zero also tell the history of our culture: each time we use the word "cipher," we are also reviving a linguistic memory of the time when a zero was a dangerous secret that could have got you burned at the stake.

Discovery of zero and the place-value system: 458 CE. (To this day, no document has been found to prove that the nine units were used at this date according to the place-value system.) The names of the first nine numbers are used according to the place-value system, as we shall see in the Lokavibhaga, dated 458 CE, where the names of the numbers are sometimes replaced by word–symbols and the word shunya (void) and its synonyms are used as zeros.

5.33 Documents showing use of the place-value system and zero

The use of place-value system and zero began to appear frequently in documents from India and Southeast Asia (the following list is nonexhaustive):

594 CE. Sankheda charter on copper,
628 CE. Brahmasputa Siddhanta (written 3 AD) by Brahmagupta,
629 CE. Commentary on the Aryabhatiya by Bhaskara,
683 CE. Khmer inscription of Trapeang Prei,
683 CE. Malaysian inscription of Kedukan Bukit,
684 CE. Malaysian inscription of Talang Tuwo,
686 CE. Malaysian inscription of Kota Kapur,
737 CE. Charter of Dhiniki on copper,
753 CE. Inscriptions of Devendravarmana,
760 CE. Javanese inscription of Dinaya,
793 CE. Charter of Rashtrakuta on copper,
813 CE. Cham inscription of Po Nagar,
815 CE. Charter of Buchkala on copper,
829 CE. Cham inscription of Bakul,
837 CE. Inscription of Bauka,

850 CE. Ganita Sara Samgraha of Mahaviracharya,
862 CE. Inscription of Deogarh,
875 CE. Inscription of Gwalior,
877 CE. Balinese inscription of Haliwanghang,
878 CE. Balinese inscription of Mamali,
880 CE. Balinese inscription of Taragal, and
917 CE. Charter on copper of Mahipala.

5.34 Ananta and Bindu in Hinduism

The word "Ananta" literally means "infinity." In Hindu mythology, the ananta denotes a huge serpent representing eternity and the immensity of space. It is shown resting on the primordial waters of original chaos. Vishnu is lying on the serpent, between two creations of the world, floating on the "ocean of unconsciousness." The serpent is always represented as coiled up, in a sort of figure eight on its side (like the symbol ∞), and theoretically has a thousand heads. It is considered to be the great king of the nagas and lord of hell (patala). Each time the serpent opens its mouth it produces an earthquake because there is a belief that the serpent also supported the world on its back. It is the serpent that at the end of each kalpa, spits the destructive fire over the whole of creation.

Ananta. Value = zero, infinity. It seems paradoxical, yet this symbolism comes from the association of Ananta, the serpent of infinity, with the immensity of space. As "space = 0," the name of the serpent became a synonym of zero.

Bindu. Value = 0, this word literally means "point." This is the symbol of the universe in its nonmanifest form, before its transformation into the world of appearances (rupadhatu). The comparison between the uncreated universe and the point is due to the fact that this is the most elementary mathematical symbol of all, yet it is capable of generating all possible lines and shapes (rupa). Thus the association of ideas with "zero," which is not only considered to be the most negligible quantity, but also and above all it is the most fundamental of mathematical concepts and the basis for all abstract sciences.

5.35 Working without concept of zero: Deterrent for Babylonians

For more than fifteen centuries, Babylonian mathematicians and astronomers worked without a concept of or sign for zero, and that must have hampered them a great deal.

5.36 Mystery continues and so does our quest

Physics—the science of matter and not of spirit/consciousness—continues to grope in the dark in search of the origin of the universe. Has the universe which we live in cropped up straight from a perfect (exact) infinite eternal void? This question implies that something has been created out of nothing and consequently this goes against the very principle of physics that we know of. Or, has it been there eternally in some form perceived/not perceived by scientists/astronomers due perhaps to either too short a time of existence of a particle (lifetime of 125 ps, say) in the process of transformation or too long a time, such as the astronomical figure like 10^{102} years, as mentioned earlier? In fact both the time scales are completely beyond any measuring device that we have. A direct measurement is completely out of question while some kind of physical arguments would provide the aforementioned periods whose numerical quality would possibly remain meaningfully unknown.

None of the models/theories such as the Big Bang Theory and the Steady State Theory proposed so far has succeeded to explain all perceived phenomena occurring in the universe. Physics alone appears to be insufficient to solve the mystery. Should we then include consciousness along with physics to arrive at a model which would explain all possible happenings in the universe? Or, only in the realm of deep meditation of the sages (spiritual scientists) of the world, can this question be answered insisting anybody to come to that state of deep meditation and get the answer in his own mind and convince himself (and perhaps not others who have not come to that mental state)? Most of the revelations/solutions to queries are known to come to a human being through intense concentration/meditation. This is how the scientists have brought the humanity to this current level in which we live in.

Bibliography

Agarwal, M. K. (2012). *From Bharata to India: Chrysee the Golden.* iUniverse.206 ISBN: 9781475907650.

Agarwal, R. P., Agarwal, H., & Sen, S. K. (2013). Birth, growth and computation of Pi to ten trillion digits. *Advances in Difference Equations, 100*, 1–59. <http://www.advancesindifferenceequations.com/content/2013/1/100>.

Agarwal, R. P., & Sen, S. K. (2014). *Creators of Mathematical and Computational Sciences.* Springer.

Algebra with Arithmetic of Brahmagupta and Bhaskara. (1817). (H. T. Colebrooke, Trans. to English). London.

Allison, G. (2013). *"Zero Dark Thirty" has the facts wrong—and that's a problem, not just for the Oscars.* The Christian Science Monitor.

Aryabhatiya of Aryabhata (W. E. Clark, Trans.).

Asimov, I. (1978). Article "Nothing Counts". In *Asimov on Numbers.* Pocket Books.

Auburn, D. (2001). *Proof: A Play.* London: Faber and Faber.

"Aught" definition, Dictionary.com Retrieved April 2013.

"Aught" synonyms, Thesaurus.com Retrieved April 2013.

Backus, J. (1977). *The Acoustical Foundations of Music* (2nd ed.). New York, NY: W.W. Norton and Co.

Barrow, J. D. (2001). *The Book of Nothing.* Vintage. ISBN: 0-09-928845-1.

Berner, R. W. (1967). Towards standards for handwritten zero and oh: Much ado about nothing (and a letter), or a partial dossier on distinguishing between handwritten zero and oh. *Communications of the ACM, 10*(8), 513–518. http://dx.doi.org/10.1145/363534.363563

Bill Casselman (University of British Columbia). (2000). American Mathematical Society. "All for Nought" G. Ifrah (Ed.), p. 400.

Binary Numbers in Ancient India. <http://home.ica.net/~roymanju/Binary.htm>.

Bourbaki, N. (1998). *Elements of the History of Mathematics* (Vol. 46, ISBN: 3-540-64767-8). Berlin, Heidelberg, and New York: Springer-Verlag.

Brezina, C. (2006). *Al-Khwarizmi: The Inventor Of Algebra.* The Rosen Publishing Group. ISBN 978-1-4042-0513-0.

Britannica *Concise Encyclopedia* (2007). Entry *algebra.*

Bunt, L. N. H., Jones, P. S., & Bedient, J. D. (1988). *The historical roots of elementary mathematics.* Courier Dover Publications. 254–255; ISBN: 0-486-2556-3.

Calinger, R. (1999). *A Conceptual History of Mathematics.* Upper Saddle River, NJ.

Chanda, M., & Sen, S. K. (1968). *Significant spiritual events in the life of Swami Vivekananda.* Bangalore: Deccan Herald. (Sunday, Jan 21, magazine Section).

Chandra Sekhar, J., & Gangadhar Prasad, M. (Eds.), (2013). *Eternally talented India: 108 facts, Vivekananda Institute of Human Excellence.* Hyderabad, India: Ramakrishna Math.

Chevalier, J. (1982). *Dictionnaire des symboles.* R. Laffont (ed.), ISBN: 2-221-50319-8.

Chibisov, G. V. (1976). Astrophysical upper limits on the photon rest mass. *Soviet Physics Uspekhi, 19*, 624.

Chisholm, H. (Ed.), (1911). 'Zero.' *Encyclopædia Britannica* (11th ed.). Cambridge University Press.

Chuquet, N. (1484). *Triparty en la science des nombres*. Unpublished in his life time; see <http://en.wikipedia.org/wiki/Nicolas_Chuquet>.

cipher | cypher, n. OED Online. December 2011. Oxford University Press. Accessed 04.03.12. Archived from the original on 2012-03-06.

Clifton, M. (1997). Star, bright. In C. Fadiman (Ed.), *Mathematical Magpie* (pp. 70–96). New York, NY: Copernicus.

Coates, R. M. (1997). The law. In C. Fadiman (Ed.), *Mathematical Magpie* (pp. 15–19). New York, NY: Copernicus.

Cody, W. J. (1988). Floating point standards—Theory and practice. In R. E. Moore (Ed.), *Reliability in Computing: The Role of Interval Methods on Scientific Computing*. Boston, MA: Academic Press.

Cody, W. J., Coonen, J. T., Gay, D. M., Hanson, K., Hough, D., & Kahan, W., et al. (1984). A proposed radix and word-length standard for floating point arithmetic. *IEEE Micro*, *4*(4), 86–100.

Coe, M. D. (1992). *Breaking the Maya code*. London.

Consciousness. Merriam-Webster. Retrieved 04.06.12.

Culbert, P. T., & Sabloff, J. A. (1995). *Maya civilisation*. New York, NY.

D'Ambrosio, U. (1993). Mathematics and literature. In A. White (Ed.), *Essays in Humanistic Mathematics*. Washington, DC: Mathematical Association of America.

Datta, B. (1931). Early literary evidence of the use of the zero in India. *The American Mathematical Monthly*, *38*(10), 566–572. <http://www.jstor.org/stable/2301384>.

Datta, B. (1932). On Mahavira's solution of rational triangles and quadrilaterals. *Bulletin of Calcutta Mathematical Society*, *20*, 267–294.

Datta, B., Singh, A. N., & Narayan, A. (1962). *History of Hindu mathematics: A Source book; Volumes I [Numeral notation and Arithmetic] and II [Algebra]* (Also, Bharatiya Kala Prakashan, 2004, Reprint, xxv, 575pp., 2 Vols, ISBN: 8186050868). Bombay: Asia Publishing House. Also, Bharatiya Kala Prakashan, 2004, Reprint, xxv, 575pp., 2 Vols, ISBN: 8186050868.

Dey, S. K. (1997). Analysis of consciousness in Vedanta philosophy. *Informatica*, *21*(3), 405–419.

Diehl, R. A. (2004). *The Olmecs: America's First Civilization*. London: Thames & Hudson.

Dijkstra, E. W. *Why numbering should start at zero*. EWD831 (PDF of a handwritten manuscript). <http://www.cs.utexas.edu/users/EWD/ewd08xx/EWD831.PDF>.

Dodgson, C. L. (1960). In M. Gardner (Ed.), *The Annotated Alice*. New York, NY: Bramhall.

Farthing, G. (1992). *The Psychology of Consciousness*. Prentice Hall. ISBN 978-0-13-728668-3.

Filliozat, J. (1957–1964). La science indienne antique. In R. Taton (Ed.), *Histoire générale des sciences* Vol. 159. Paris.

Fins, J. J., Schiff, N. D., & Foley, K. M. (2007). Late recovery from the minimally conscious state: Ethical and policy implications. *Neurology*, *68*(4), 304–307.

Fischbach, E., Kloor, H., Langel, R. A., Lui, A. T. Y., & Peredo, M. (1994). New geomagnetic limits on the photon mass and on long range forces coexisting with electromagnetism. *Physical Review Letters*, *73*, 514–517.

Forsythe, G. E., & Moler, C. B. (1967). *Computer solution of Linear Algebraic Systems*. Englewood Cliffs, NJ: Prentice-Hall.

Godel, K. (1931). Uber formal unedtscheidhare Satze der Principa Mathematica and verwandter Systeme, I. *Monatshefte fur Mathematik und Physik*, *38*, 173–198.

Grattan-Guinness, I. (1997). *The Fontana History of the Mathematical Sciences*. Fontana Press.

Gray, L. H. (1913). *Vasavadatta of Subandhu (A Sanskrit Romance)*. 1965 reprint: ISBN: 978-0-404-50478-6; 1999 reprint: ISBN: 81-208-1675-7.

Gray, L. H. (1999). *Subandhu's Vāsavadattā: A Sanskrit Romance*. Delhi: Motilal Banarsidass.

Grimm, R. E. (1973). The autobiography of Leonardo Pisano. *Fibonacci Quarterly*, *11*(1), 99–104.

Güzeldere, G. (1997). In N. Block, O. Flanagan & G. Güzeldere (Eds.), *The Nature of Consciousness: Philosophical debates* (pp. 1–67). Cambridge, MA: MIT Press.

Hameroff, S., Kaszniak, A., & Chalmers, D. (1999). Preface. In *Toward a Science of Consciousness III: The Third Tucson Discussions and Debates*. MIT Press.xixxx ISBN: 978-0-262-58181-3.

Hayashi, T. (1992). Mahavira's formulas for a conch-like plane figure. *Ganita Bharati*, *14*(1–4), 1–10.

Hindu Vedic philosophy (Hinduism, Philosophy, Science and History). <http://hinduismphilosophysciencehistory.blogspot.in/2013/07/articles-research-on-hinduism-2.html>.

Hodgkin, L. (2005). *A History of Mathematics: From Mesopotamia to Modernity*. Oxford University Press. 85 ISBN: 978-0-19-152383-0.

Hyslop, A. (1995). *Other Minds*. Springer. 5–14; ISBN: 978-0-7923-3245-9.

Ifrah, G. (1987). (L. Bair, Trans.) *From One to Zero: A Universal History of Numbers*. New York, NY: Penguin.

Ifrah, G. (2000). p. 416.

Jain, A. (1984). Mahaviracarya, the man and the mathematician. *Acta Ciencia Indica Mathematics*, *10*(4), 275–280.

Jeans, J. (1968). *Science and Music*. Dover.154.

Joseph, G. G. (2011). *The Crest of the Peacock: Non-European Roots of Mathematics* (3rd ed.) ISBN: 978-0-691-13526-7. Princeton. 86.

Kanigel, R. (1992). *The Man Who Knew Infinity: A Life of the Genius Ramanujan*. New York, NY: Washington Square Press.

Kaplan, R., & Kaplan, E. (2000). *The Nothing That Is: A Natural History of Zero*. Oxford: Oxford University Press.

Keith, A. B. (1993). *A History of Sanskrit Literature*. Delhi: Motilal Banarsidass. ISBN: 81-208-1100-3.

Knobe, J. (2008). *Can a Robot, an Insect or God Be Aware?* Scientific American: Mind.

Krishnamurthy, E. V., & Sen, S. K. (2009). *Numerical Algorithms: Computations in Science and Engineering*. New Delhi: Affiliated East West Press.

Kuroda, M., Michiwaki, H., Saitoh, S., & Yamane, M. (2014). New meanings of the division by zero and interpretations on 100/0=0 and on 0/0=0. *International Journal of Applied Mathematics*, *27*(2), 191–198.

Lakshmikantham, V., & Devi, J. V. (2006). *What India Should Know*. Mumbai: Bharatiya Vidya Bhavan.

Lakshmikantham, V., Leela, S., & Devi, J. V. (2005). *The Origin and History of Mathematics*. Cambridge, UK: Cambridge Scientific Publishers.

Lakshmikantham, V., & Sen, S. K. (2005). *Computational Error and Complexity in Science and Engineering*. Amsterdam: Elsevier.

Lemma B.2.2, (1999). The integer 0 is even and is not odd. In R. C. Penner (Ed.), *Discrete Mathematics: Proof Techniques and Mathematical Structures* (pp. 34). World Scientific. ISBN: 981-02-4088-0.

Mahendra Nath Dutta (younger brother of Swami Vivekananda), Vivekananda Swamijir Jiboner Ghatanaboli (Bengali), Part 1, Mohendra Publishing Committee, Kolkata, (3rd ed.), 1965 (Bengali year 1371); (*The English translation of the title is "Events in the life of Swami Vivekananda" first edition was published in 1938 i.e. Bengali year 1332*).

Makemson, M. W. (1946). *The Maya correlation problem*. Poughkeepsie, NY.

Mandler, G. (1975). Consciousness: Respectable, useful, and probably necessary. In R. Solso (Ed.), *Information processing and cognition: The Loyola symposium (Also in: Technical Report No. 41, Center for Human Information Processing, University of California, San Diego. March, 1974* (pp. 229–254). Hillsdale, NJ: Lawrence Erlbaum Associates.

Mandler, G. (2002). *Consciousness recovered: Psychological functions and origins of thought*. Philadelphia, PA: John Benjamins.

Marlow, A. R. (Ed.), (1980). *Quantum theory and gravitation*. New York, NY: Academic Press. (Proceedings of a symposium held at Loyola University, New Orleans, May 23–26, 1979).

Mathematics *in the near and far east*. p. 262.

Menninger, K. (1992). *Number words and number symbols: A cultural history of numbers*. Courier Dover Publications.401 ISBN: 0-486-27096-3.

Midgley, M. (2001). *Science and Poetry*. London: Routledge.

Mukherjee, R. (1991). *Discovery of zero and its impact on Indian mathematics*. Calcutta.

Part XXIII. Reprinted Mathematics in literature Newman, J. R. (Ed.). (2000). *The World of Mathematics* (Vol. IV, pp. 2214–2277). New York, NY: Dover.

No conspiracy: New documents explain Pentagon, CIA cooperation on "Zero Dark Thirty". (2012). *Entertainment Weekly*.

O'Connor, J. J., & Robertson, E. F. (2013). *Aryabhata the Elder*. Scotland: School of Mathematics and Statistics University of St Andrews. Retrieved 26.05.13.

Pandit, M. D. (1993). *Mathematics as known to the Vedic Samhitas*. New Delhi: Sri Satguru Publications. 298–299.

Pannekoek, A. (1961). *A History of Astronomy*. George Allen & Unwin.165.

Pierce, J. R. (1983). *The Science of Musical Sound*. New York, NY: Scientific American Books, Inc.

Pogliani, L., Randic, M., & Trianjstic, N. (1998). Much ado about nothing—An introductive inquiry about zero. *International Journal of Mathematical Education in Science and Technology*, *29*(5), 729–744.

Ranade, D. (2008). Bose-Einstein condensate and State of Samadhi. *Times of India* (Dec 30).

Reid, C. (1992). *From zero to infinity: What makes numbers interesting* (4th ed.) ISBN: 978-0-88385-505-8. Mathematical Association of America. 23.

Robert Temple. *The Genius of China, A place for zero*. ISBN: 1-85375-292-4.

Rossing, T. D. (1982). *The Science of Sound*. Reading, MA: Addison-Wesley Publishing Co.

Rouse Ball, W. W. (1888). *A Short Account of the History of Mathematics*. Dover Publications. <http://store.doverpublications.com/0486206300.html>.

Russell, B. (1942). *Principles of mathematics* (2nd ed.) ISBN: 1-4400-5416-9, (Chapter 14). Forgotten Books.125.

Sabloff, J. A. (1990). *The New Archaeology and the Ancient Maya*. London.

Salomon, R. (1995). On the origin of the early indian scripts: A review article. *Journal of the American Oriental Society*, *115*(2), 271–279.

Salomon, R. (1996). Brahmi and Kharoshthi. In P. T. Daniels & W. Bright (Eds.), *The World's Writing Systems*. Oxford University Press. ISBN: 0-19-507993-0.

Salomon, R. (1998). *Indian Epigraphy: A Guide to the Study of Inscriptions in Sanskrit, Prakrit, and the Other Indo-Aryan Languages*. Oxford: Oxford University Press. ISBN: 0-19-509984-2.

Sanchez, G. I. (1961). *Arithmetic in Maya*. Texas.

Satyam's Raju: From small spinning unit to spinning big lies. **Deccan Herald** (Daily newspaper, Bangalore edition, Apr 10). (2015).

Schneider, S., & Velmans, M. (2008). Introduction. In *Max Velmans, Susan Schneider. The Blackwell Companion to Consciousness*. Wiley. ISBN: 978-0-470-75145-9.

Searle, J. (2005). Consciousness. In T. Honderich (Ed.), *The Oxford companion to philosophy*. Oxford University Press. ISBN: 978-0-19-926479-7.

Seife, C. (2000). *Zero: The Biography of a Dangerous Idea*. USA: Penguin. (Paper). ISBN: 0-14-029647-6.

Sen, S. K. (2003). Error and computational complexity in engineering. In J. C. Misra (Ed.), *Computational Mathematics, Modelling and Algorithms*. New Delhi: Narosa Pub. House.

Sen, S. K. (2014). Natural mathematics, computer mathematics, and mathematics: Scope in engineering computation. *Nonlinear Studies, 21*(2), 309–318.

Sen, S. K. (2014). Extraordinary mental abilities of Swami Vivekananda: A scientific explanation. *Vedanta Kesari*, May issue, 35–38.

Sen, S. K., & Agarwal, R. P. (2011). *Pi, e, phi with Matlab: Random and Rational Sequences with Scope in Supercomputing Era*. UK: Cambridge Scientific Publishers.

Shen, K. -S., Liu, H., & Lun, A. W. C. (1999). *The Nine Chapters on the Mathematical Art: Companion and Commentary*. Oxford University Press. 35. ISBN: 978-0-19-853936-0. "zero was regarded as a number in India… whereas the Chinese employed a vacant position".

Sigler, L. (2003). (English translation) *Fibonacci's Liber Abaci*. Springer.

Sivaram, C. (2014). *Still in the dark*. Bangalore: Deccan Herald, Aug 05.

Soanes, C., Waite, M., & Hawker, S. (Eds.), (2001). *The Oxford Dictionary, Thesaurus and Wordpower Guide (Hardback)*. New York, NY: Oxford University Press. ISBN: 978-0-19-860373-3.

Srinivasachariar, T. V. (1906). *Vasavadatta of Subandhu*. Trichinopoly: St. Joseph's College Press.

Sri Vidyaranya Swami. (1967). Pancadasi: A comprehensive technical manual (based on pure reasoning) written in the Sanskrit Sloka format over 700 years ago (English translation and notes by Swami Swahananda), Advaita Ashrama.

Stahl, W. H. (1962). *Roman Science*. Madison, WI: University of Wisconsin Press.

Steel, D. (2000). *Marking time: The epic quest to invent the perfect calendar*. John Wiley & Sons.113 ISBN: 0-471-29827-1. "In the B.C./A.D. scheme there is no year zero. After 31 December 1 BC came AD 1 January 1. …If you object to that no-year-zero scheme, then don't use it: Use the astronomer's counting scheme, with negative year numbers."

Struik, D. J. (1987). *A Concise History of Mathematics*. New York, NY: Dover Publications; 32–33. In these matrices we find negative numbers, which appear here for the first time in history

Swami Gambhirananda (Ed.). (1899–1988) 11th President Ramakrishna order. *A Short Biography of Swami Vivekananda*.

Swami Sarvapriyananda, The Eternal Witness, Youtube, Swamiji is a monk of Ramakrishna order and an expert in Vedanta philosophy, who speaks in the language of modern time. <https://www.youtube.com/watch?v=CzB5k1RhFqE>.

Teresi, D. (1997). Zero. *The Atlantic Monthly*, July, 88.

Thakurdesai, M. A. (2012). *Higgs they trust (Heart of the matter)*. Bangalore: Deccan Herald, Jan 30.

The Britannica *Guide to Numbers and Measurement (Math Explained)*. 2010. The Rosen Publishing Group. ISBN: 9781615301089. (pp. 97–98).

Trattati d'aritmetica pubblicati da Baldassarre Boncompagni, I, Algoritmi de numero Indorum; II, Ioannis Hispalensis liber Algoritmi de practica arismetice. Roma. (1857).

Tryon, E. P. (1973). Is the Universe Vacuum Fluctuation? *Nature, 246*, 396–97.

Two Mathematical Offerings, Association of Mathematics Teachers of India (2013).

van Gulick, R. (2004). Consciousness. Stanford Encyclopedia of Philosophy.

Van Nooten, B. (1993). Binary numbers in Indian antiquity. *Journal of Indian Studies*, *21*, 31–50.

Vilenkin, A. (1994). Quantum Cosmology and the Initial State of the Universe. *Physical Review D*, *50*, 2581–94.

Wallin, N. -B. (2012). How was zero discovered? *YaleGlobal (A publication of the Macmillan Center)* <http://yaleglobal.yale.edu/about/zero.jsp>.

Whitehead, A. N., & Russell, B. (1910–1913). *Principia mathematica, 1 (1910), 2(1912), 3 (1913)*. London: Cambridge University Press.

Zeleznika, A. P. (1997). Informational theory of consciousness. *Informatica*, *21*(3), 345–369.

Zero. (1920). *Encyclopedia Americana*.

Index

Note : Page numbers followed by "*f*" and "*t*" refer to figures and tables, respectively.

D

E

F

G

T

U

V

W

Y

Z

Printed in the United States
By Bookmasters